AGRICULTURE ISSUES AND POLICIES

TRANSFORMATION OF U.S. ANIMAL AGRICULTURE

Agriculture Issues and Policies

Assessing Disease Potential in U.S. Aquaculture Industry
Emilija Kalnins (Editor)
2009. ISBN: 978-1-60741-543-5

Agricultural Conservation
Anthony G. Hargis (Editor)
2009. ISBN: 978-1-60692-273-6
2009. ISBN: 978-1-60876-813-4 (E-book)

Environmental Services and Agriculture
Karl T. Poston (Editor)
2009. ISBN: 978-1-60741-053-9

Weeds: Management, Economic Impacts and Biology
Rudolph V. Kingely (Editor)
2009. ISBN: 978-1-60741-010-2

No-Till Farming: Effects on Soil, Pros and Cons and Potential
Earl T. Nardali (Editor)
2009. ISBN: 978-1-60741-402-5

Inspection and Protection of U.S. Meat and Poultry
Finn J. Amundson (Editor)
2009. ISBN: 978-1-60741-120-8

Global Beef Trade
Alessandro Ferrara (Editor)
2009. ISBN: 978-1-60741-121-5

Organic Food - Economics and Issues
Earl D. Straub (Editor)
2009. ISBN: 978-1-60741-130-7

Soybean and Wheat Crops: Growth, Fertilization and Yield
Samuel Davies and George Evans
2009. ISBN: 978-1-60741-173-4
2009. ISBN: 978-1-61668-984-1 (E-book)

Effects of Liberalizing World Agricultural Trade
Henrik J. Ehrstrom (Editor)
2009. ISBN: 978-1-60741-198-7
2009. ISBN: 978-1-60876-601-7 (E-book)

Hired Farmworkers: Profile and Labor Issues
Rea S. Berube (Editor)
2009. ISBN: 978-1-60741-232-8

Agricultural Wastes
Geoffrey S. Ashworth and Pablo Azevedo (Editors)
2009. ISBN: 978-1-60741-305-9

Sugar Beet Crops: Growth, Fertilization & Yield
Claus T. Hertsburg (Editor)
2009. ISBN: 978-1-60741-491-9

Economic Impacts of Foreign-Source Animal Disease
Jace R. Corder (Editor)
2009. ISBN: 978-1-60741-601-2
2009. ISBN: 978-1-60876-602-4 (E-book)

Essential Oils: Art, Agriculture, Science, Industry and Entrepreneurship (A Focus on the Asia-Pacific Region)
Murray Hunter
2009. ISBN: 978-1-60741-865-8

Drivers and Restraints for Economically Efficient Farm Production
Helena Hansson Karin Larsén Bo Öhlmér
2009. ISBN: 978-1-60876-171-5

Agriculture and Food
Claus Schäfer
2010. ISBN: 978-1-60692-038-1

Corn Crop Production: Growth, Fertilization and Yield
Arn T. Danforth (Editor)
2009. ISBN: 978-1-60741-955-6
2009. ISBN: 978-1-60876-860-8 (E-book)

Phosphate Solubilizing Microbes for Crop Improvement
Mohammad Saghir Khan and Almas Zaidi (Editors)
2009. ISBN: 978-1-60876-112-8

Agricultural Economics: New Research
Tomas H. Lee (Editor)
2010. ISBN: 978-1-61668-077-0
2010. ISBN: 978-1-61668-445-7 (E-book)

U.S. Biobased Products Market Potential and Projections through 2025
Meredith A. Williamson (Editor)
2010 ISBN: 978-1-60741-033-1

Genetically Engineered Crops and Proposed Oversight
Brandon C. Vaughn (Editor)
2010 ISBN: 978-1-60876-067-1
2010 ISBN: 978-1-61728-094-8 (E-book)

The Sugar Industry and Cotton Crops
Peter T. Jenkins (Editor)
2010. ISBN: 978-1-61668-320-7
2010. ISBN: 978-1-61668-439-6 (E-book)

Tomatoes: Agricultural Procedures, Pathogen Interactions and Health Effects
Eric D. Aubé and Frederick H. Poole (Editors)
2010. ISBN: 978-1-60876-869-1

Sustainable Agriculture: Technology, Planning and Management
Augusto Salazar and Ismael Rios (Editors)
2010. ISBN: 978-1-60876-269-9

Organic Farming and Peanut Crops
Darren C. Grossman and Terrance L. Barrios (Editors)
2010. ISBN: 978-1-60876-187-6

Farm Bill of 2008: Major Provisions and Legislative Action
Mary T. Conner (Editor)
2010. ISBN: 978-1-60741-750-7

America's Family Farms
Efren J. Tamayo (Editor)
2010. ISBN: 978-1-60741-751-4

Million Dollar Farms in the New Century
Samuel D. Bosworth (Editor)
2010. ISBN: 978-1-60741-755-2
2010. ISBN: 978-1-61668-541-6 (E-book)

Pesticide Resistance, Population Dynamics and Invasive Species Management
Gregory J. McKee, Colln A. Carter, James A. Chalfant, Rachael E. Goodhue, and Frank G. Zalom
2010. ISBN: 978-1-60741-758-3

Agroforestry Development on the Canadian Prairies
Suren N. Kulshreshtha, Ken Van Rees, Hayley Hesseln, Mark Johnston, John Kort
2010. ISBN: 978-1-61668-266-8
2010. ISBN: 978-1-61668-588-1 (E-book)

Agriculture Research and Technology
Kristian Bundgaard and Luke Isaksen (Editors)
2010 ISBN: 978-1-60741-850-4

Agricultural Production
Felix C. Wager (Editor)
2010. ISBN: 978-1-61668-695-6
2010. ISBN: 978-1-61728-050-4 (E-book)

Ecophysiology of Tropical Tree Crops
Fabio DeMatta (Editor)
2010. ISBN: 978-1-60876-392-4

Rethinking Structural Reform in Turkish Agriculture: Beyond the World Bank's Strategy
Baris Karapinar, Fikret Adaman and Gokhan Ozertan (Editors)
2010. ISBN: 978-1-60876-718-2

Desalination: Methods, Costs and Technology
Irena A. Urbonienė (Editor)
2010. ISBN: 978-1-61668-909-4
2010. ISBN: 978-1-61728-242-3 (E-book)

Manure Use for Fertilizer and Energy
Connor D. Macias (Editor)
2010. ISBN: 978-1-60876-847-9

Price Dynamics behind Consumer Food Purchases
Morgan D. Fitzpatrick (Editor)
2010. ISBN: 978-1-60876-892-9

Governance of Agrarian Sustainability
Hrabrin Bachev
2010. ISBN: 978-1-60876-888-2

Transformation of U.S. Animal Agriculture
Justin M. Daigle (Editor)
2010. ISBN: 978-1-60876-938-4

The Peanut Plant and Light: Spermidines from Peanut Flowers
Victor S. Sobolev, James B. Gloe and Arlene A. Sy
2010. ISBN: 978-1-61668-028-2
2010. ISBN: 978-1-61668-371-9 (E-book)

Soil Phenols
A. Muscolo and M. Sidari (Editors)
2010. ISBN: 978-1-60876-264-4

Agriculture and Environmental Security in Southern Ontario's Watersheds
Glen Filson, Bamidele Adekunle, and Katia Marzall
2010. ISBN: 978-1-61668-156-2
2010. ISBN: 978-1-61668-372-6 (E-book)

Alignment-Free Models in Plant Genomics: Theoretical, Experimental, and Legal issues
Humberto González-Díaz, Guillermin Agüero-Chapin, Cristian Robert Munteanu, Francisco Prado-Prado, Kuo-Chen Chou, Aliuska Duardo-Sanchez, Grace Patlewicz and Antonio López-Diaz
2010. ISBN: 978-61668-333-7
2010. ISBN: 978-61668-603-1 (E-book)

Manure: Management, Uses and Environmental Impacts
Carmen S. Dellaguardia (Editor)
2010. ISBN: 978-1-61668-424-2
2010. ISBN: 978-1-61668-647-5 (E-book)

Postharvest Management of Litchi and Rambutan Fruits
D. Sivakumar and L. Korsten
2010. ISBN: 978-1-61668-688-8
2010. ISBN: 978-1-61728-089-4 (E-book)

Understanding U.S. Farm Exits, Farm Survival and the Role of Commodity Payments
Lucas A. Hamilton (Editor)
2010. ISBN: 978-1-61668-995-7
2010. ISBN: 978-1-61728-245-4 (E-book)

AGRICULTURE ISSUES AND POLICIES

TRANSFORMATION OF U.S. ANIMAL AGRICULTURE

JUSTIN M. DAIGLE
EDITOR

Nova Science Publishers, Inc.

New York

For permission to use material from this book please contact us:
Telephone 631-231-7269; Fax 631-231-8175
Web Site: http://www.novapublishers.com

LIBRARY OF CONGRESS CATALOGING-IN-PUBLICATION DATA

Transformation of U.S. animal agriculture / editor, Justin M. Daigle.
p. cm.
Includes index.
ISBN 978-1-60876-938-4 (hardcover)
1. Livestock systems--United States. 2. Animal waste--Law and legislation--United States. 3. Livestock--United States--Genetics. 4. Swine--Economic aspects--United States. 5. Dairying--Economic aspects--United States. 6. Broilers (Poultry)--Economic aspects--United States. I. Daigle, Justin M.
SF51.T72 2010
338.1'760973--dc22

2009051586

Published by Nova Science Publishers, Inc. † New York

CONTENTS

PREFACE

The animal sector of agriculture has undergone major changes in the last several decades, a fact that has drawn the attention of policymakers and the public. In particular, organizational changes within the industry to enhance economic efficiency have resulted in larger confined production facilities that often are geographically concentrated. Increased facility size and regional concentration of livestock and poultry operations have, in turn, given rise to concerns over the management of animal wastes from these facilities and potential impacts on environmental quality, public health and welfare. This book reviews some of the current laws and legislative issues regarding animal agriculture. Moreover, the authors of this book discuss the application of biotechnology to improve productivity, consistency, and quality; to introduce new food, fiber and medical products; and to protect the environment. This book relies on farm-level data to detail the nature, causes, and effects of structural changes in livestock production.

Chapter 1 - The animal sector of agriculture has undergone major changes in the last several decades: organizational changes within the industry to enhance economic efficiency have resulted in larger confined production facilities that often are geographically concentrated. These changes, in turn, have given rise to concerns over the management of animal wastes and potential impacts on environmental quality.

Federal environmental law does not regulate all agricultural activities, but certain large animal feeding operations (AFOs) where animals are housed and raised in confinement are subject to regulation. The issue of applicability of these laws to livestock and poultry operations— especially the Comprehensive Environmental Response, Compensation, and Liability Act (CERCLA, the Superfund law) and the Emergency Planning and Community Right-to-Know Act (EPCRA)—has been controversial and recently has drawn congressional attention.

Both Superfund and EPCRA have reporting requirements that are triggered when specified quantities of certain substances are released to the environment. In addition, Superfund authorizes federal cleanup of releases of hazardous substances, pollutants, or contaminants and imposes strict liability for cleanup and injuries to natural resources from releases of hazardous substances.

Superfund and EPCRA include citizen suit provisions that have been used to sue poultry producers and swine operations for violations of those laws. In two cases, environmental advocates claimed that AFO operators had failed to report ammonia emissions, in violation of Superfund and EPCRA. In both cases, federal courts supported broad interpretation of key terms defining applicability of the laws' reporting requirements. Three other cases not dealing with reporting violations also have attracted attention, in part because of questions of whether animal wastes contain hazardous substances that can create cleanup and natural resource damage liability under Superfund. Two of these cases were settled; the third, brought by the Oklahoma Attorney General against poultry operations in Arkansas, is pending.

These lawsuits testing the applicability of Superfund and EPCRA to poultry and livestock operations have led to congressional interest in these issues. In the 110th Congress, legislation was introduced that would have amended CERCLA to clarify that manure is not a hazardous substance, pollutant, or contaminant under that act and that the laws' notification requirements would not apply to releases of manure (H.R. 1398 and S. 807). Proponents argue that Congress did not intend that either of these laws apply to agriculture and that enforcement and regulatory mechanisms under other laws are adequate to address environmental releases from animal agriculture. Opponents respond that enacting an exemption would severely hamper the ability of government and citizens to know about and respond to releases of hazardous substances caused by an animal agriculture operation. In December 2008, EPA issued a rule to exempt animal waste emissions to the air from most CERCLA and EPCRA reporting requirements.

Chapter 2 - Animal agriculture is being transformed by rapid advances in biotechnology—a term that encompasses a variety of technologies, including genetic engineering (GE), genetic modification, transgenics, recombinant DNA techniques, and cloning, among others. Producers are interested in the application of biotechnology to improve productivity, consistency, and quality; to introduce new food, fiber, and medical products; and to protect the environment. Potential human health applications of transgenic animals include producing biopharmaceuticals and generating organs, tissues, and cells for xenotransplantation. Criticisms of such applications involve issues ranging from food safety and social resistance to potential negative impacts on animal welfare and on ecosystems. Questions also have arisen about the adequacy of the current regulatory structure to assess and manage any risks created by these technologies.

On January 15, 2009, the U.S. Food and Drug Administration (FDA) released final guidance on how it is to regulate GE animals and products. FDA is to do so under its existing statutory authority and regulations. Generally, GE-derived foods, for example, are to be regulated like non- GE foods; if their composition does not differ from their conventional counterparts, they will not have to be labeled. Nonetheless, developers of GE animals and of GE-derived products must gain FDA pre-market approval.

On February 6, 2009, the FDA announced the first approval of a drug from a GE animal. The drug is a human anti-clotting agent produced in the milk of transgenic goats.

Although animal biotechnology involves many techniques other than cloning, this latter technology has attracted widespread attention. A final risk assessment and industry guidance on the safety of meat and milk from cloned cattle, pigs, and goats and their offspring were released January 15, 2008, by FDA. The documents generally echoed the FDA's December 28, 2006, draft risk assessment, which found that such products are as safe to eat as those of conventionally bred animals. The FDA also concluded that cloning poses the same risks to animal health as those found in animals created through other assisted reproductive technologies—although the frequency of such problems is higher in cloning. (Scientists stress that cloning is an assisted reproduction technique that does not involve any transfer or alteration of genes through GE.) The agency said it was no longer asking industry to refrain voluntarily from marketing the products of cloned animals and their offspring, although the U.S. Department of Agriculture (USDA) did ask that it be continued for products from clones (but not from the offspring of clones).

The Senate-passed version of the 2007-2008 farm bill would have delayed the FDA final risk assessment and continued the marketing moratorium until completion of newly mandated studies on the safety and on the market impacts of introducing such products. This delay was not retained in the enacted version of the bill (P.L. 110-234). Bills on animal cloning in the 110th Congress would have required all food from cloned animals or their offspring to be labeled, and prohibited food from cloned animals from being labeled as organic. These and other bills on cloning or other regulation of animal biotechnology could be offered in the 111th Congress.

Chapter 3 - The increasing size and specialization of hog operations reflect structural change in U.S. swine production during the past 15 years. The number of farms with hogs has declined by over 70 percent, as hog enterprises have grown larger. Large operations that specialize in a single phase of production have replaced farrow-to-finish operations that performed all phases of production. The use of production contracts has increased. Operations producing under contract are larger than independent operations and are more likely to specialize in a single phase of production. These structural changes have coincided with substantial gains in efficiency for hog farms and lower production costs. Most of these productivity gains are attributable to increases in the scale of production and technological innovation. Productivity gains likely contributed to a 30-percent reduction in the price of hogs at the farm gate.

Chapter 4 - U.S. dairy production is consolidating into fewer but larger farms. This report uses data from several USDA surveys to detail that consolidation and to analyze the financial drivers of consolidation. Specifically, larger farms realize lower production costs. Although small dairy farms realize higher revenue per hundredweight of milk sold, the cost advantages of larger size allow large farms to be profitable, on average, even while most small farms are unable to earn enough to replace their capital. Further survey evidence, as well as the financial data, suggest that consolidation is likely to continue.

Chapter 5 - Broiler production in the United States is coordinated almost entirely through systems of production contracts, in which a grower's compensation is based, in part, on how the grower's performance compares with that of other growers. The industry is undergoing a gradual structural change as production shifts to larger broiler enterprises that provide larger shares of an operator's household income. Larger enterprises require substantially larger investments in broiler housing, and new or retrofi tted houses are also an important source of productivity growth in the industry. This report, based on a large and representative survey of broiler operations, describes the industry's organization, housing features, contract design, fees and enterprise cost structures, and farm and household finances.

Chapter 6 - U.S. livestock production has shifted to much larger and more specialized farms, and the various stages of input provision, farm production, and processing are now much more tightly coordinated through formal contracts and shared ownership of assets. Important financial advantages have driven these structural changes, which in turn have boosted productivity growth in the livestock sector. But structural changes can also generate environmental and health risks for society, as industrialization concentrates animals and animal wastes in localized areas. This report relies on farm-level data to detail the nature, causes, and effects of structural changes in livestock production.

In: Transformation of U.S. Animal Agriculture
Editor: Justin M. Daigle

ISBN: 978-1-60876-938-4

Chapter 1

ANIMAL WASTE AND HAZARDOUS SUBSTANCES: CURRENT LAWS AND LEGISLATIVE ISSUES

Claudia Copeland
Resources and Environmental Policy, (CRS)

SUMMARY

The animal sector of agriculture has undergone major changes in the last several decades: organizational changes within the industry to enhance economic efficiency have resulted in larger confined production facilities that often are geographically concentrated. These changes, in turn, have given rise to concerns over the management of animal wastes and potential impacts on environmental quality.

Federal environmental law does not regulate all agricultural activities, but certain large animal feeding operations (AFOs) where animals are housed and raised in confinement are subject to regulation. The issue of applicability of these laws to livestock and poultry operations— especially the Comprehensive Environmental Response, Compensation, and Liability Act (CERCLA, the Superfund law) and the Emergency Planning and Community Right-to-Know Act (EPCRA)—has been controversial and recently has drawn congressional attention.

Both Superfund and EPCRA have reporting requirements that are triggered when specified quantities of certain substances are released to the environment. In addition, Superfund authorizes federal cleanup of releases of hazardous substances, pollutants, or contaminants and imposes strict liability for cleanup and injuries to natural resources from releases of hazardous substances.

Superfund and EPCRA include citizen suit provisions that have been used to sue poultry producers and swine operations for violations of those laws. In two cases, environmental advocates claimed that AFO operators had failed to report ammonia emissions, in violation of Superfund and EPCRA. In both cases, federal courts supported broad interpretation of key terms defining applicability of the laws' reporting requirements. Three other cases not dealing with reporting violations also have attracted attention, in part because of questions of whether animal wastes contain hazardous substances that can create cleanup and natural resource damage liability under Superfund. Two of these cases were settled; the third, brought by the Oklahoma Attorney General against poultry operations in Arkansas, is pending.

These lawsuits testing the applicability of Superfund and EPCRA to poultry and livestock operations have led to congressional interest in these issues. In the 110th Congress, legislation was introduced that would have amended CERCLA to clarify that manure is not a hazardous substance, pollutant, or contaminant under that act and that the laws' notification requirements would not apply to releases of manure (H.R. 1398 and S. 807). Proponents argue that Congress did not intend that either of these laws apply to agriculture and that enforcement and regulatory mechanisms under other laws are adequate to address environmental releases from animal agriculture. Opponents respond that enacting an exemption would severely hamper the ability of government and citizens to know about and respond to releases of hazardous substances caused by an animal agriculture operation. In December 2008, EPA issued a rule to exempt animal waste emissions to the air from most CERCLA and EPCRA reporting requirements.

INTRODUCTION

The animal sector of agriculture has undergone major changes in the last several decades, a fact that has drawn the attention of policymakers and the public. In particular, organizational changes within the industry to enhance economic efficiency have resulted in larger confined production facilities that often are geographically concentrated.[1] Increased facility size and regional concentration of livestock and poultry operations have, in turn, given rise to concerns over the management of animal wastes from these facilities and potential impacts on environmental quality, public health and welfare.

Animal manure can be and frequently is used beneficially on farms to fertilize crops and add or restore nutrients to soil. However, animal waste, if not properly managed, can adversely impact water quality through surface runoff and erosion, direct discharges to surface waters, spills and other dry-weather discharges, and leaching into soil and ground. It can also result in emission to the air of particles and gases such as ammonia, hydrogen sulfide, and volatile organic chemicals. According to the U.S. Department of Agriculture (USDA), in 1997, 66,000 operations had farm- level excess nitrogen (an imbalance between the quantity of manure nutrients produced on the farm and assimilative capacity of the soil on that farm), and 89,000 had farm-level excess phosphorus.[2] USDA believes that where manure nutrients exceed the assimilative capacity of a region, the potential is high for runoff, leaching of nutrients, and other environmental problems. Geographically, areas with excess farm-level nutrients correspond to areas with increasing numbers of confined animals.

Federal environmental law does not regulate all agricultural activities. Some laws specifically exempt agriculture from regulatory provisions, and others are structured so that farms escape most, if not all, of the regulatory impact. Still, certain large animal feeding operations (AFOs) where animals are kept and raised in confinement are subject to environmental regulation. The primary regulatory focus on environmental impacts has been on protecting water resources and has occurred under the Clean Water Act. In addition, facilities that emit large quantities of air pollutants may be regulated under the Clean Air Act.

Some livestock operations also may be subject to requirements of the Comprehensive Environmental Response, Compensation, and Liability Act (CERCLA, the Superfund law) and the Emergency Planning and Community Rightto-Know Act (EPCRA).[3] The issue of applicability of these laws to livestock and poultry operations—especially CERCLA and EPCRA—has been controversial and has drawn congressional attention.

This report describes the provisions of Superfund and EPCRA, and enforcement actions under these laws that have increasingly been receiving attention. Congressional scrutiny in the form of legislative proposals and two House hearings are discussed. Bills intended to exempt animal manure from the requirements of Superfund and EPCRA were introduced in the 109th Congress. Similar bills were introduced in the 110th Congress (H.R. 1398 and S. 807), but no legislation has been enacted. Issues raised by the legislation are analyzed.

CERCLA AND EPCRA

Both the Comprehensive Environmental Response, Compensation, and Liability Act (the Superfund law, 42 U.S.C. §§9601-9675) and the Emergency Planning and Community Right-toKnow Act (42 U.S.C. § § 11001-11050) have reporting requirements that are triggered when specified quantities of certain substances are released to the environment.[4] Both laws, which are administered by the Environmental Protection Agency (EPA), utilize information disclosure in order to increase the information available to government and citizens about the sources and magnitude of chemical releases to the environment. In addition to reporting requirements, CERCLA includes provisions authorizing federal cleanup of releases of hazardous substances, pollutants, or contaminants that may present an imminent and substantial danger to the public health or welfare (Section 104), and imposing strict liability for cleanup and damages for injury to, destruction of, or loss of natural resources resulting from releases of hazardous substances (Section 107). At issue today is how the reporting requirements and other provisions of these laws apply to poultry and livestock operations.

Superfund authorizes programs to remediate uncontrolled or abandoned hazardous waste sites and assigns liability for the associated costs of cleanup. Section 103(a) of CERCLA requires that the person in charge of a facility (as defined in Section 101(9)) that releases a "reportable quantity" of certain hazardous substances must provide notification of the release to the National Response Center.

EPCRA establishes requirements for emergency planning and notification for storage and release of hazardous and toxic chemicals. Section 304(a)(1) of EPCRA requires the owner or operator of a facility (as defined in Section 329(4)) to report to state and local authorities any releases greater than the reportable quantity of substances deemed hazardous under Superfund or extremely hazardous under EPCRA. Under Superfund, the term "release" (Section 10 1(22)) includes discharges of substances to water and land and emissions to the air from "spilling, leaking, pumping, pouring, emitting, emptying, discharging, injection, escaping, leaching, dumping, or disposing into the environment." Under EPCRA, the term "release" (Section 329(8)) includes emitting any hazardous chemical or extremely hazardous substance into the environment. Superfund excludes the "normal application of fertilizer" from the definition of release, and EPCRA excludes from the definition of hazardous chemicals any substance that is "used in routine agricultural operations or is a fertilizer held for sale by a retailer to the ultimate customer."

The CERCLA definition of "hazardous substance" (Section 101(14)) triggers reporting under both laws. Among the reportable substances that may be released by livestock facilities are hydrogen sulfide, ammonia, and phosphorus. The reportable quantity (RQ) for both hydrogen sulfide and ammonia is 100 pounds per day, or 18.3 tons per year; the RQ for phosphorus is 1 pound per day. Section 109 of Superfund and Section 325 of EPCRA authorize EPA to assess civil penalties for failure to report releases of hazardous substances that equal or exceed their reportable quantities (up to $27,500 per day under CERCLA and $27,500 per violation under EPCRA).

ENFORCEMENT AGAINST AFOS

EPA has enforced the Superfund and EPCRA reporting requirements against AFO release of hazardous pollutants in two separate cases. The first involved the nation's second largest pork producer, Premium Standard Farms (PSF) and Continental Grain Company. In November 2001, EPA and the Department of Justice announced an agreement resolving numerous claims against PSF concerning principally the Clean Water Act, but also the Clean Air Act, Superfund, and EPCRA. More recently, in September 2006, the Department announced settlement of claims against Seaboard Foods—a large pork producer with more than 200 farms in Oklahoma, Kansas, Texas, and Colorado—and PIC USA, the former owner and operator of several Oklahoma farms now operated by Seaboard. Like the earlier PSF case, the government had brought complaints for violations of several environmental laws, including failure to comply with the release reporting requirements of CERCLA and EPCRA.

Both Superfund and EPCRA include citizen suit provisions that have been used to sue poultry producers and swine operations for violations of the laws (Section 310 of CERCLA and Section 326 of EPCRA). In two cases, environmental advocates claimed that AFO operators had failed to report ammonia emissions, putting them in violation of Superfund and EPCRA. In both cases, federal courts supported broad interpretation of key terms defining applicability of the laws' reporting requirements.

In the first of these cases, a federal district court in Oklahoma initially ruled in 2002 that a farm's individual barns, lagoons, and land application areas are separate "facilities" for purposes of CERCLA reporting requirements, rather than aggregating multiple emissions of pollutants across the entire site. This court held that Superfund's reporting requirements would only apply if emissions for *each individual facility* exceed 100 pounds per day. However, the district court's ruling was reversed on appeal (*Sierra Club v. Seaboard Farms Inc.,* 387 F.3d 1167 (10th Cir. 2004)). The court of appeals ruled that the whole farm site is the proper entity to be assessed for purposes of CERCLA reporting and determining if emissions of covered hazardous substances meet minimum thresholds.

In the second case, a federal district court in Kentucky similarly ruled in 2003 that the term "facility" should be interpreted broadly to include facilities operated together for a single purpose at one site, and that the whole farm site is the proper entity to be assessed for purposes of the Superfund and EPCRA reporting requirements (*Sierra Club v. Tyson Foods, Inc.,* 299 F.Supp. 2d 693 (W.D. Ky. 2003)). While Superfund provides that a continuous release is subject to reduced reporting requirements, and EPCRA provides an exemption for reporting releases when the covered substance is used in routine agricultural operations or is used on other farms for fertilizer, the court found that these exemptions did not apply to the facts of this case. The ruling was not appealed.

EPA was not a party in either of these lawsuits. The U.S. Court of Appeals for the 10th Circuit invited EPA to file an amicus brief in the *Seaboard Farms* case in order to clarify the government's position on the issues, but EPA declined to do so within the time frame specified by the court.

Three other cases in federal courts, while they do not include reporting violations, also have drawn attention, in part because they raised the question of whether animal wastes that contain phosphorus are hazardous substances that can create cleanup and natural resource injury liability under Superfund.[5] Animal wastes typically contain low levels of phosphorus, and animal wastes are beneficially used as fertilizer on farms. Over the long term, however, the application of animal waste fertilizers may result in phosphorus buildup in soils which may be released to watersheds through surface runoff. In 2003, a federal court in Oklahoma held that phosphorus contained in poultry litter in the form of phosphate is a hazardous substance under Superfund and thus could subject poultry litter releases to provisions of that law (*City of Tulsa v. Tyson Foods, Inc.*, 258 F. Supp. 2d 1263, (N.D. Okla. 2003)). This ruling was later vacated as part of a settlement agreement, but some observers believe that the court's reasoning may still be persuasive with other courts. The second case, *City of Waco v. Schouten* (W.D. Tex., No. W-04- CA- 118, filed April 29, 2004), was brought by the city against 14 dairies alleging various causes of action based on disposal of wastes from those operations. It was resolved by a settlement agreement early in 2006.[6]

The third case, *State of Oklahoma v. Tyson Foods, Inc.* (N.D. Okla, No. 4:05-cv-00329, filed June 13, 2005), is still pending. This suit, brought by the Oklahoma Attorney General, asserts various claims based on the disposal of waste from 14 poultry operations in the Illinois River Watershed. The state principally seeks its past and present response costs and natural resource injuries under CERCLA due to release of wastes from these facilities.

The net result of these lawsuits is growing concern by the agriculture community that other legal actions will be brought and that the courts will continue to hold that the Superfund and EPCRA reporting requirements and other provisions apply to whole farm sites, thus potentially exposing more of these operations to enforcement under federal law.

Reporting Exemption

In 2005, a group of poultry producers petitioned EPA for an exemption from EPCRA and CERCLA emergency notification requirements for releases of ammonia, arguing that such releases from poultry growing operations pose little or no risk to public health, while reporting imposes an undue burden on the regulated community and government responders.[7] In 2007, EPA formed an internal workgroup to review information on animal waste as it relates to CERCLA and to possible exemptions from emissions reporting, and in February, EPA Administrator Stephen Johnson told congressional committees that the agency would propose a rule to exempt routine animal waste air releases from emergency notification requirements.

In December 2007, EPA issued a proposal in response to the poultry industry petition.[8] The proposal would exempt releases of hazardous substances to the air (typically during digestion or decomposition) from animal waste at farms from the notification requirements of CERCLA and EPCRA. The exemption would apply to releases to the air from manure, digestive emissions, and urea, including animal waste mixed with bedding, compost, and other specified materials. "Farm" is defined in the proposal as an agricultural operation from which $1,000 or more of agricultural products are sold annually (the same definition used by the Department of Agriculture). EPA explained that the rule is justified because of the resource burden to industry of complying with reporting requirements, since the agency cannot foresee a situation where a response action would be taken as a result of notification of releases of hazardous substances from animal waste at farms.[9]

The proposal drew significant public response during the comment period. While the proposed exemption pleased many agriculture industry groups who seek a waiver or other means to limit possible liability under CERCLA and EPCRA, environmental advocates and other stakeholders opposed the exemption, saying that emissions from animal wastes are not trivial or benign. Critics noted that the EPA proposal would exempt releases of ammonia, as originally requested in the industry petition, plus hydrogen sulfide and all other hazardous chemicals, such as nitrous oxide and volatile organic compounds released from animal waste. Some argued that an exemption is premature, since EPA is moving forward with research on emissions levels, which could be undermined by a regulatory exemption (see CRS Report RL3 2947, *Air Quality Issues and Animal Agriculture: EPA's Air Compliance Agreement*). State air quality officials opposed blanket regulatory or legislative exemptions, and they recommended that if the agency considers any action, it should only be a narrow exemption, such as one based on a size threshold for farms.[10]

In December 2008, EPA finalized the CERCLA/EPCRA administrative reporting exemption with some modifications to the original proposal.[11] The final rule exempts hazardous substance releases that are emitted to the air from animal waste at farms from the notification requirement of CERCLA. As proposed, the final rule relieves all livestock operations of all size, not just poultry farms, from CERCLA's requirement to report hazardous substance releases to the air to federal officials. In addition, the final rule provides a partial exemption for such releases from EPCRA's requirement to report releases to state and local emergency officials. Partially responding to public comments, the final rule continues to apply EPCRA's reporting requirement to large animal feeding operations (those that are subject to permitting requirements under the Clean Water Act[12]), but exempts smaller facilities. A number of groups criticized the final rule, again raising concern about the toxicity of chemicals such as ammonia and hydrogen sulfide that are emitted from animal waste facilities and arguing the CERCLA and EPCRA do not authorize administrative exemptions for specific industries. A coalition of environmental advocates challenged the rule in federal court (Waterkeeper Alliance v. EPA, D.C. Cir., No. 09-1017, January 15, 2009).

In September 2008, the Government Accountability Office (GAO) issued a report evaluating EPA's activities to regulate air emissions and water discharges from animal feeding operations. GAO found that EPA is unable to assess the extent to which pollution from feedlots may be impairing human health and the environment, because it lacks data on the amount of pollutants that CAFOs are releasing to the air and water. GAO recommended that EPA develop a comprehensive national inventory of CWA-permitted CAFOs and accelerate its efforts to develop protocols for measuring and quantifying air contaminants from animal feedlots. GAO noted that EPA has been criticized because its current air emissions monitoring activities are limited in scope and sample size and may not produce sufficient information to shape future regulation. Moreover, GAO questioned the basis for the CERCLA/EPCRA exemption that EPA proposed in 2007. "It is unclear how EPA made this determination when it has not yet completed its data collection effort and does not yet know the extent to which animal feeding operations are emitting these pollutants."[13]

CONGRESSIONAL INTEREST

The court cases testing the applicability of Superfund and EPCRA to poultry and livestock operations have led to congressional interest in these issues. In March 2004, a number of senators wrote to the EPA Administrator to ask the agency to clarify the reporting requirements of the two laws so as to limit their impact on poultry operations. The senators' letter said that because of unclear regulations and a lack of scientific information about emissions, poultry and livestock producers are uncertain about the laws' requirements and are vulnerable to enforcement actions.[14] In report language accompanying EPA's FY2006 appropriations, the House Appropriations Committee urged EPA to address the issues.

> The Committee continues to be concerned that unclear regulations, conflicting court decisions, and inadequate scientific information are creating confusion about the extent to which reporting requirements in [CERCLA] and [EPCRA] cover emissions from poultry, dairy, or livestock operations. Producers want to meet their environmental obligations but need clarification from the Environmental Protection Agency on whether these laws apply to their operations. The committee believes that an expeditious resolution of this matter is warranted.[15]

Also in 2004, some in Congress considered proposing legislation that would amend the definition of "release" in Superfund (Section 101(22); 42 U.S.C. §9601(22)) to clarify that the reporting requirements do not apply to releases from biological processes in agricultural operations and to amend EPCRA to exclude releases of hazardous chemicals produced through biological processes in routine agricultural operations. For some time, there were indications that an amendment containing these statutory changes would be offered during debate on FY2005 consolidated appropriations legislation, but this did not occur.[16]

Some Members sought to amend the FY2006 Agriculture appropriations bill, H.R. 2744, with a provision exempting releases of livestock manure from CERCLA and EPCRA. The proposal was promoted by Senate conferees on the bill, but it was not accepted by House conferees. Proponents, including Senator Larry Craig, contended that the proposed language was consistent with current law, because in their view CERCLA and EPCRA were never intended to apply to agriculture. Environmentalists objected to the language, arguing that it could prevent public health authorities from responding to hazardous substance releases from AFOs, would block citizen suits against agriculture companies for violations of reporting requirements, and would create an exemption from Superfund liability for natural resource injuries that might result from a large manure spill. EPA's congressional affairs office released an unofficial analysis criticizing the bill. It argued that, by eliminating federal liability for manure releases under Superfund and EPCRA, the provision could interfere with

EPA's Air Compliance Agreement, because companies would have much less incentive to participate in the agreement. The agreement is a plan that EPA announced in January 2005 to collect air quality monitoring data on animal agriculture emissions.[17] The House and Senate gave final approval to H.R. 2744 in November 2005 (P.L. 109-97), without the language that Senate conferees had proposed.

Also in November 2005, legislation was introduced that would amend CERCLA to clarify that manure is not a hazardous substance, pollutant, or contaminant under that act and that CERCLA's notification requirements would not apply to releases of manure (H.R. 4341). The bill was similar to the legislative language that Senator Craig had proposed to conferees as a provision of the FY2006 Agriculture appropriations bill with a broad definition of "manure" that includes, for example, bedding commingled with animal waste.

H.R. 4341 was introduced the same day that a House Energy and Commerce subcommittee held a hearing on animal agriculture and Superfund. The Subcommittee on Environment and Hazardous Materials heard from agriculture industry witnesses who urged Congress to provide policy direction on the issue that has developed as a result of recent and potential litigation. Other witnesses testified that the reporting and notification requirements of Superfund and EPCRA provide a safety net for making information on releases available to government and citizens, and that other environmental laws, such as the Clean Air Act, cannot function in that manner. An EPA witness said that the agency was at the time considering ways to reduce the paperwork burdens for large AFOs to report their emissions (resulting in the December 2007 proposed regulatory exemption, discussed above). Related legislation was introduced in the Senate (S. 3681). Similar legislation was introduced in the 110th Congress (H.R. 1398 and S. 807), but no further action occurred on any of these bills.

During consideration of farm bill legislation in 2007 (H.R. 2419), the House Agriculture Committee approved an amendment expressing a sense of the Committee that farm manure is not to be considered a toxic waste. However, the amendment was not included in the reported version of the bill, reportedly because of jurisdictional issues, and was not included in the final legislation (P.L. 110-246).

Some Members of Congress have been critical of EPA's proposal to exempt routine animal waste air releases from CERCLA and EPCRA's reporting requirements, questioning the potential for harmful environmental and enforcement impacts of the proposal.[18] At a September 24, 2008, hearing where GAO's recent report was discussed (see page), several House Energy and Commerce subcommittee members said that they are skeptical of the EPA's authority for a blanket exemption. Others suggested that an exemption for small farms, whose emissions are unlikely to cause environmental harm, would make sense. EPA and USDA witnesses supported the proposal, saying that the air release waiver would only affect reporting meant for emergency response situations, but would not affect requirements to report emissions of hazardous substances from other farm sources, or releases of hazardous substances from manure into soil, ground water, or surface water.

Policy Issues

Supporters and opponents of the 109th Congress and 110th Congress legislation raised a number of arguments for and against the proposals. For example, proponents of the exemption proposed in these bills, representing the agriculture industry, especially livestock and poultry producers, say that animal manure has been safely used as a fertilizer and soil amendment by many cultures all over the world for centuries and thus should not be considered a hazardous substance. Opponents—including environmental activists, public health advocates, and state and local governments—agree that when properly managed, manure has beneficial uses. Superfund's reporting and cost recovery requirements do not threaten responsible operators who manage manure as a valuable fertilizer, they say. However, these groups say that when improperly managed and in the massive amounts produced at today's large feedlot operations, animal waste can release a number of polluting substances to the environment. Releases to surface water, groundwater, and the atmosphere may include nutrients, organic matter, solids, pathogens, volatile compounds, particulate matter, antibiotics, pesticides, hormones, gases that are associated with climate change (carbon dioxide and methane), and odor.

Proponents of the legislation argue that neither Superfund nor EPCRA was intended by Congress to apply to agriculture and that the pending legislation would simply clarify congressional intent. CERCLA exempts "normal application of fertilizer" from the definition of "release" and also exempts releases of "naturally occurring organic substances." Animal waste arguably was intended to be covered by these existing exemptions, they say. Opponents respond that there is little firm evidence either way on this point, as there is limited legislative history concerning this language. The exemption for "normal application of fertilizer," enacted in CERCLA in 1980, applies to application of fertilizer on crops or cropland for beneficial use, but does not mean dumping or disposal of larger amounts or concentrations than are beneficial to crops.[19]

EPA has not issued guidance to interpret what constitutes "normal application of fertilizer," and the only court decision so far addressing this issue (the vacated 2003 *City of Tulsa* case discussed above) held that neither plaintiffs nor defendants in that case had presented evidence sufficient for a fact-based determination of what constitutes "normal application." Opponents of the legislation also argue that animal manure consists of a number of substances that are nutritional and pharmaceutical elements of the feed provided to animals (trace elements, antibiotics, nutrients), and releases are the result of inadequate waste disposal, not "naturally occurring" substances and activities.

Proponents argue that enforcement and regulatory mechanisms exist under the Clean Water Act (CWA) and other media-specific statutes, such as the Clean Air Act (CAA), making it unnecessary to rely on Superfund or EPCRA for enforcement or remediation. In particular, both the Clean Water Act and Clean Air Act require that regulated facilities obtain permits that authorize discharges or emissions of pollutants. Enforcement of permit requirements has been an important tool for government and citizens to address environmental concerns of animal agriculture activities.

Opponents respond that enforcement under Superfund fills critical gaps in these other environmental laws, because not all pollutants are covered by other laws. For example, releases of ammonia and hydrogen sulfide are listed under CERCLA but are not currently regulated as hazardous pollutants under the CAA. Clean Water Act AFO permits primarily address discharges of nutrients, but not other components of manure waste (e.g., trace elements, metals, pesticides, pathogens). Moreover, neither of these laws provides for recovery of costs for responding to or remediating releases, nor for natural resource injuries. Opponents also argue that, while "federally permitted releases" are exempt from CERCLA's reporting requirements, CWA and CAA permit requirements apply only to facilities that meet specified regulatory thresholds (for example, CWA permit rules apply to about 14,000 large AFOs, less than 6% of all AFOs in the United States).[20]

Finally, proponents of the legislation argue that if animal manure is considered to be a hazardous substance under Superfund, farm operations both large and small potentially could be exposed to costly liabilities and penalties. Opponents note that the purpose of release reporting is to keep federal, state, and local entities informed and to alert appropriate first responders of emergencies that might necessitate response, such as release of hazardous chemicals that could endanger public health in a community. The exemption proposed in pending legislation, they point out, would apply not only to CERCLA and EPCRA reporting requirements but also to other provisions (such as Superfund's authority for federal cleanup of releases, cleanup liability, and liability for natural resource injuries).

According to states and some other interest groups, liability, which arises when manure is applied in amounts that exceed what is beneficial to support crops, is necessary to bring about improvements in waste handling practices of large AFOs. Enacting an exemption would severely hamper the ability of government to appropriately respond to releases of hazardous substances and pollution caused by an animal agriculture operation, they argue. On the issue of penalties, opponents note that penalties are not available under Superfund for removal or remedial actions (except for failure to comply with information gathering and access related to a response action), regardless of whether initiated by government or a private party. CERCLA does authorize civil penalties for violation of the Section 103 reporting requirements (up to $27,500 per day), but neither of the two key citizen suit cases decided thus far (*Sierra Club v. Tyson Foods, Inc., and Sierra Club v. Seaboard Farms Inc.*) involved penalties for failure to report releases.

CONCLUSION

Issues concerning the applicability of Superfund and EPCRA to animal agriculture activities have been controversial and have drawn considerable attention. Bills in the 109th Congress gained much support (in the 109th Congress, H.R. 4341 had 191 co-sponsors, and S. 3681 had 35 cosponsors), but were not enacted. They also drew opposition from environmental advocacy groups and state and local governments. The Bush Administration did not present an official position on the legislation. Continuing interest in the issue is evident from the fact that similar legislation has been introduced in the 110th Congress.

End Notes

[1] For additional information, see CRS Report RL33325, *Livestock Marketing and Competition Issues*, by Renée Johnson and Geoffrey S. Becker.

[2] U.S. Department of Agriculture, Natural Resources Conservation Service, "Manure Nutrients Relative to the Capacity of Cropland and Pastureland to Assimilate Nutrients: Spatial and Temporal Trends for the United States," Publication no. nps00-579, December 2000, p. 85.

[3] For additional information, see CRS Report RL32948, *Air Quality Issues and Animal Agriculture: A Primer*, by Claudia Copeland.

[4] For additional information on Superfund and EPCRA, see CRS Report RL30798, *Environmental Laws: Summaries of Major Statutes Administered by the Environmental Protection Agency (EPA)*, by Susan R. Fletcher et al., and CRS Report RL33426, *Superfund: Implementation and Selected Issues*, by Mark Reisch and Jonathan L. Ramseur.

[5] Unlike the citizen suit cases discussed above, these lawsuits do not address what is a "facility," for purposes of determining whether a release has occurred. EPA also is not a party in any of these cases.

[6] In July 2008, a federal appeals court ruled in another case that animal waste discharges cannot be challenged under the Resource Conservation and Recovery Act (RCRA) if the discharges are already regulated under a Clean Water Act (CWA) permit. RCRA establishes a regulatory scheme for the treatment, disposal, and storage of hazardous and solid wastes. The court ruled that RCRA does not apply to any activity or substance that is subject to the CWA. Coon v. Willet Dairy, No. 07-3454-cv (CA 2, July 30, 2008).

[7] In 1998, EPA granted an administrative exemption from release reporting requirements for certain radionuclide releases. EPA cited authority in CERCLA sections 102(a), 103, and 115 for granting administrative reporting exemptions where "releases of hazardous substances that pose little or no risk or to which a Federal response is infeasible or inappropriate." See 63 *Federal Register* 13461 (March 19, 1998).

[8] U.S. Environmental Protection Agency, "CERCLA/EPCRA Administration Reporting Exemption for Air Releases of Hazardous Substances from Animal Waste," 72 *Federal Register* 73700 (December 28, 2007).

[9] Ibid., p. 73704.

[10] National Association of Clean Air Agencies, letter to the Honorable Barbara Boxer, chairman, Senate Environment and Public Works Committee, March 20, 2007.

[11] U.S. Environmental Protection Agency, "CERCLA/EPCRA Administrative Reporting Exemption for Air Releases," 73 *Federal Register* 76948-76960, December 18, 2008.

[12] For additional information, see CRS Report RL3 1851, *Animal Waste and Water Quality: EPA Regulation of Concentrated Animal Feeding Operations (CAFOs)*, by Claudia Copeland.

[13] U.S. Government Accountability Office, "Concentrated Animal Feeding Operations, EPA Needs More Information and a Clearly Defined Strategy to Protect Air and Water Quality from Pollutants of Concern," September 2008, GAO- 08-944, p. 7.

[14] Senator Blanche L. Lincoln et al., letter to Michael Leavitt, EPA Administrator, March 12, 2004.

[15] U.S. Congress, House Committee on Appropriations, *Report accompanying H.R. 2361, Department of the Interior, Environment, and Related Agencies Appropriation Bill, 2006*, H.Rept. 109-80, 109th Cong., 1st sess., p. 87.

[16] "Spending Bill Excludes Proposal for Farms; Craig Plans Separate Legislation Next Year," *Daily Environment Report*, November 23, 2004, p. A-10.

[17] For information, see CRS Report RL32947, *Air Quality Issues and Animal Agriculture: EPA's Air Compliance Agreement*, coordinated by Claudia Copeland.

[18] Letter from Reps. John Dingell, Albert Wynn, Hilda Solis to Stephen L. Johnson, EPA Administrator, March 18, 2008.

[19] U.S. Senate, Committee on Environment and Public Works, *Environmental Emergency Response Act, Report to Accompany S. 1480*, 96th Cong., 2nd sess., S.Rept. 96-848, p. 46.

[20] U.S. Environmental Protection Agency, "National Pollutant Discharge Elimination System Permit Regulation and Effluent Limitation Guidelines and Standards for Concentrated Animal Feeding Operations (CAFOs); Final Rule," 68 *Federal Register* 7179, February 12, 2003.

In: Transformation of U.S. Animal Agriculture
Editor: Justin M. Daigle
ISBN: 978-1-60876-938-4

Chapter 2

BIOTECHNOLOGY IN ANIMAL AGRICULTURE: STATUS AND CURRENT ISSUES

S. Geoffrey Becker[1] and Tadlock Cowan[2]
[1]Resources and Rural Development, (CRS)
[2]Natural Resources and Rural Development

SUMMARY

Animal agriculture is being transformed by rapid advances in biotechnology—a term that encompasses a variety of technologies, including genetic engineering (GE), genetic modification, transgenics, recombinant DNA techniques, and cloning, among others. Producers are interested in the application of biotechnology to improve productivity, consistency, and quality; to introduce new food, fiber, and medical products; and to protect the environment. Potential human health applications of transgenic animals include producing biopharmaceuticals and generating organs, tissues, and cells for xenotransplantation. Criticisms of such applications involve issues ranging from food safety and social resistance to potential negative impacts on animal welfare and on ecosystems. Questions also have arisen about the adequacy of the current regulatory structure to assess and manage any risks created by these technologies.

On January 15, 2009, the U.S. Food and Drug Administration (FDA) released final guidance on how it is to regulate GE animals and products. FDA is to do so under its existing statutory authority and regulations. Generally, GE-derived foods, for example, are to be regulated like non- GE foods; if their composition does not differ from their conventional counterparts, they will not have to be labeled. Nonetheless, developers of GE animals and of GE-derived products must gain FDA pre-market approval.

On February 6, 2009, the FDA announced the first approval of a drug from a GE animal. The drug is a human anti-clotting agent produced in the milk of transgenic goats.

Although animal biotechnology involves many techniques other than cloning, this latter technology has attracted widespread attention. A final risk assessment and industry guidance on the safety of meat and milk from cloned cattle, pigs, and goats and their offspring were released January 15, 2008, by FDA. The documents generally echoed the FDA's December 28, 2006, draft risk assessment, which found that such products are as safe to eat as those of conventionally bred animals. The FDA also concluded that cloning poses the same risks to animal health as those found in animals created through other assisted reproductive technologies—although the frequency of such problems is higher in cloning. (Scientists stress that cloning is an assisted reproduction technique that does not involve any transfer or alteration of genes through GE.) The agency said it was no longer asking industry to refrain voluntarily from marketing the products of cloned animals and their offspring, although the U.S. Department of Agriculture (USDA) did ask that it be continued for products from clones (but not from the offspring of clones).

The Senate-passed version of the 2007-2008 farm bill would have delayed the FDA final risk assessment and continued the marketing moratorium until completion of newly mandated studies on the safety and on the market impacts of introducing such products. This delay was not retained in the enacted version of the bill (P.L. 110-234). Bills on animal cloning in the 110th Congress would have required all food from cloned animals or their offspring to be labeled, and prohibited food from cloned animals from being labeled as organic. These and other bills on cloning or other regulation of animal biotechnology could be offered in the 111th Congress.

INTRODUCTION

The U.S. Food and Drug Administration (FDA) released two recent documents that renewed public and congressional interest in animal biotechnology in general and animal cloning in particular. On January 15, 2009, the agency released final guidance representing its current thinking on the regulation of genetically engineered (GE) animals for food or drugs. FDA did so under its existing statutory authority and regulations.[1] A year earlier, on January 15, 2008, FDA had unveiled its final risk assessment and industry guidance on the safety of milk and meat from cloned animals and their offspring. The FDA found that, generally, meat and milk products from cattle, pigs, and goats are as safe as products from their non-cloned counterparts.[2]

Still, some in the United States remain concerned about the safety and other impacts of GE animals and of animal cloning, and they continue to advocate a cautious approach to commercialization of such products. Outside of the United States, animal biotechnology, including both GE and cloning, is the focus of ongoing regulatory, policy and scientific discussions within the European Union (EU) as well as within the international organization for animal health, known by its French acronym, the OIE. (Cloning by itself is not considered to be genetic engineering; see discussion later in this report).

Biotechnology is a broadly defined term of relatively recent origin describing the range of modern knowledge, applications, and techniques underlying advances in many fields, notably health care and agriculture. Animal biotechnology has been defined as "that set of techniques by which living creatures are modified for the benefit of humans and other animals."[3] By its very nature, agricultural development is the history of humans modifying plants and animals to maximize desirable traits. For example, domestication and selective breeding of animals date back many thousands of years. Artificial insemination of livestock, notably dairy cattle, is a more recent technology, first finding wide commercial acceptance in the 1950s.

Discovery of the genetic code in the 1950s gave birth to modern techniques of biotechnology. One of the first commercial products of this new biotechnology in animal agriculture was bovine somatotropin (bST), a naturally occurring metabolic modifier that is now being manufactured in larger quantities through the use of recombinant DNA technology. Manufactured bST came onto the market in 1994 and is now administered to as many as half of all U.S. dairy cattle to increase per-cow milk output. Although bST is being used commercially in approximately 20 countries, it is banned in the European Union (EU).

The first U.S. approval of a commercial product from a GE animal is for the drug ATryn, an anti- clotting agent being produced in the milk of transgenic goats. The FDA announced its approval on February 6, 2009.

Other developments include pigs that have been engineered for increased sow milk output to produce faster-growing piglets. Cloned cattle also have been developed to resist mastitis, and transgenic salmon with enhanced growth characteristics are under regulatory consideration for possible commercialization. Output traits such as drugs recovered from animal milk ("pharming"), milk that lacks allergenic proteins, and animal organs for human transplant (xenotransplantation) that resist rejection are other contemporary objectives of animal biotechnology research. In March 2006, researchers at the University of Missouri announced the creation of transgenic pigs whose tissue contains omega-3 fatty acids.[4] The consumption of omega-3 fatty acids, found primarily in fish, has been linked to lowered incidence of heart disease in humans. Similar research is also under way to produce omega-3 fatty acids in cow's milk and in chicken eggs.

This report describes several scientifically emerging animal biotechnologies that are raising a variety of questions concerning risks to humans, animals, and the environment, as well as ethical concerns. The report examines applications of the technologies and discusses major issues that may arise. Consumers, agricultural producers, the biotechnology industry, and federal regulatory bodies are debating the relative costs and benefits of these technologies. As technologies move toward commercialization, Congress is being asked to examine these issues and possibly to refine the current federal regulatory structure governing the technologies and their agricultural products.

ANIMAL BIOTECHNOLOGIES

Given the breadth of the term "animal biotechnology," one might reasonably define it to include thousands of years of humans selectively breeding animals: observing desirable animal traits and attempting to breed those traits into successive lines of animals. One of the first modern forms of assisted reproductive technology (ART) was artificial insemination (AI). AI has been long established as a technological advance in traditional selective breeding and an important adjunct to the development of modern industrial animal production, especially in dairy and poultry. AI was adopted by producers and accepted by the public with virtually no controversy. For example, more than 70% of all U.S.-bred Holstein cows, by far the most widely used milk producers, are artificially inseminated.[5] Estrus synchronization, which improves the efficiency of AI by more accurately controlling when a female was in heat, is also an important animal biotechnology.

With the development in the 1970s and patenting in the 1980s of recombinant DNA techniques, and the subsequent analysis of genes, their resulting proteins, and the role played by the proteins in animal biochemical processes (functional genomics), modern biotechnology is increasingly equipped with a set of sophisticated tools holding the promise of transforming the selective breeding of animals. The range of new techniques and technologies could transform animal biotechnology in ways that plant biotechnology was transformed in the 1980s and 1990s.

Modern animal biotechnology is developing against the background of public experience with plant biotechnology, and controversy over the technologies may be a continuing feature of animal biotechnology development, not least because of the closer connection between humans and some animals and the belief that techniques developed for animals are only a step away from application to humans. Some of the better known animal biotechnologies follow. A number of them are types of assisted reproductive technology (ART).

Embryo Transfer

After AI and estrus synchronization, embryo transfer (ET) is the third most commonly used biotechnology. In ET, a donor cow of superior breeding is chemically induced to superovulate. The eggs are then fertilized within the donor, the embryo develops and is then removed and implanted in a recipient cow. Between removal and implantation, embryos may be frozen for safekeeping. Because of the relatively high costs, ET is used mostly within registered cowherds.

Transgenics

A prominent area of contemporary animal biotechnology research is the development of *transgenic animals* through genetic engineering (GE) technology.[6] Transgenic animals are produced by introducing an isolated DNA fragment into an embryo so that the resulting animal will express a desired trait. Transgenic animals may be generated by the introduction of foreign DNA obtained through animals of the same species, animals of different species, microbes, humans, cells, and *in vitro* nucleic acid synthesis. The only currently routine use of transgenic animals, primarily mice, is in the area of human disease research. The anti-clotting agent in goat milk is the first such application to be approved by the FDA. Potential agricultural applications for such genetic engineering, however, include improved feed use and faster growth; more resistance to disease; meat that is leaner or that has more of some other desirable quality; and possibly even animal waste that is more environmentally benign. **Table 1** provides examples of various objectives of animal biotechnology involving genetic modification.

In Vitro Fertilization

With in vitro fertilization (IVF), a technician removes unfertilized eggs (oocytes) from the donor cow's ovaries, usually recovering 6-8 useable oocytes. The oocytes mature in an incubator and are fertilized with sperm. The resulting zygotes incubate and develop in the laboratory before being placed into the recipient cow. While IVF can produce many fertilized embryos, the added expense of ET makes the procedure prohibitive in most cases.

Sexing Embryos

The dairy industry prefers heifers and the beef industry prefers bulls. Embryo sexing methods in cattle have been developed using a bovine Y-chromosome probe. Technicians remove a few cells from the embryo and assess the DNA in these cells for the presence of a Y-chromosome. Presence of a Y-chromosome determines the embryo is male. Research is also developing in sperm sexing technology.

Table 1. Agricultural Applications of Animal Transgenics

Purpose	Animal model	Transgenic source
Faster growth/leaner meat	Cattle, swine, rabbits, sheep	Growth hormones/factors: *Human, Bovine, Porcine, Rat, Chicken*
Altered milk composition (higher protein)	Cattle	Extra copies of casein genes; disruption of lactoglobulin gene: *Cow*
Anti-clotting drug production in milk[a]	Goat	Human antithrombin gene
"Biosteel " production in milk[b]	Goat	Spider
Reduced phosphorus in swine feces	Swine	Phytase gene; *Bacteria*
Increased wool production	Sheep	Cysteine synthesis gene: *Bacteria* Growth factor: *Sheep*
Disease resistance	Swine,sheep, rabbit	Monoclonal antibodies: *Mouse* Viral envelope genes: *Sheep*
Xenotransplantation: Developing animal organs for human transplantation	Swine	CD55 (DAF-decay activating factor: *Human* CD59: *Human*

Source: GeneWatch UK, April 2002.

a. FDA approval for commercial use announced February 6, 2009. Source: FDA.

b. "Biosteel" is the trade name for spider web material intended to be produced in the milk of a transgenic goat. Said to be twenty times stronger than steel, "Biosteel" has an envisioned breaking strength of about 300,000 pounds per square inch and could produce microscopically fine, super strong fibers for industrial use.

Cloning

Cloning, discussed at greater length later in this report, is a biotechnology developing rapidly and with significant public controversy. Most people think of cloning as the creation of an organism that is genetically identical to another one. However, scientists use the term more broadly, to refer to production not only of such organisms but also of genetically identical cells, and to replication of DNA and other molecules. It also refers to a form of reproduction found naturally in many single-celled organisms, as well as plants and animals. Those differences in meaning and usage have caused some confusion in public debate about cloning, where the main area of controversy relates to artificial cloning involving higher organisms, including humans.[7]

Gene Knockout

This is a technique where researchers inactivate, or "knock out," a gene by replacing it or disrupting it with an artificial piece of DNA in order to determine what that particular gene does—e.g., cause or protect against some disease, alter metabolism, and so forth. A knockout mouse is a laboratory mouse subjected to this technology.[8]

Regulation and Oversight

The basic federal guidance for regulating the products of agricultural biotechnology is the Coordinated Framework for Regulation of Biotechnology (51 *Fed. Reg.* 23302), published in 1986 by the White House Office of Science and Technology Policy (OSTP). A key principle has been that GE products should continue to be regulated according to their characteristics and unique features, not their methods of production, that is, whether or not they were created through biotechnology. The framework provides a regulatory approach intended to ensure the safety of biotechnology research and products, using existing statutory authority and previous policy experience.

Some newer applications of biotechnology did not exist when the current regulatory framework was enunciated. The NRC animal biotechnology report concluded that this regulatory regime "might not be adequate to address unique problems and characteristics associated with animal biotechnologies" and that federal agency responsibilities are not clear.[9]

Food and Drug Administration (FDA)

Within the Department of Health and Human Services (HHS), the FDA regulates food, animal feed ingredients, and human and animal drugs, primarily under the Federal Food, Drug, and Cosmetic Act (FFDCA; 21 U.S.C. §301 *et seq.*). The FDA has stated that most—although probably not all—gene-based modifications of animals for production or therapeutic claims fall within the purview of the agency's Center for Veterinary Medicine (CVM), which regulates them under the FFDCA as new animal drugs. A new animal drug (NAD) must be approved by the agency after it is demonstrated to be safe to man and animals, as well as being effective. Regulation of transgenic animals as NADs, however, suggests to some observers (e.g., the Center for Food Safety, Union of Concerned Scientists) the inherent weakness of existing regulatory structures to respond adequately to the complexities that arise with animal biotechnology innovations.[10]

Primarily under the FFDCA, FDA's Center for Food Safety and Applied Nutrition (CFSAN) is responsible for assuring that domestic and imported foods are safe and properly labeled. Generally, FDA does not review new foods themselves for safety before they enter commerce but does have enforcement authority to act if it finds foods that are adulterated under the act. All food *additives*, whether or not introduced through biotechnology, must receive FDA safety approval before they can be sold; the exception to pre-market approval are those on a list the FDA has determined to be "generally recognized as safe" (GRAS).[11, 12]

Sections of the FFDCA and of the Public Health Service Act (42 U.S.C. §262 *et seq.*) provide the authorities for FDA's Center for Drug Evaluation and Research and Center for Biologics Evaluation and Research to regulate the safety and effectiveness of human drugs and other medical products, including those produced by GM animals. Under these laws, the FDA requires pre-market review and licensing of such products, and requires that their production conditions ensure purity and potency.[13]

FDA Approval of Drug in Goat Milk

The FDA's approval of the drug ATryn was the first of a commercially available product from a GE animal.FDA announced its decision on February 6, 2009, and published the final rule as a new animal drug application (NADA) in the February 11, 2009, *Federal Register* (74 *Fed. Reg.* 6823). The drug is an anticoagulant for the treatment of human patients with hereditary antithrombin deficiency; it is derived from the milk of goats bred with the introduction of an rDNA construct directing the expression of the human gene antithrombin.The milk is being produced by a herd of GE dairy goats in Massachusetts, which, the agency stated, had not demonstrated any negative health problems over seven generations.Controls are in place to ensure that no milk or meat from the goats will enter the food or feed supply, according to FDA. Biotechnology industry officials expressed their support for the approval; a leading animal activist group, the Humane Society of the United States, countered that the action "seems to perpetuate the notion of [animals] being merely tools for human use rather than sentient creatures.."

FDA Guidance on GE Animals

On January 15, 2009, FDA's CVM released its industry guidance on how it plans to regulate GE animals.[14] This final document hews closely (with a few modifications) to the draft version that FDA published on September 18, 2008. The final document asserts that FDA's authority to regulate GE animals comes under the new animal drug provisions of the FFDCA. A drug is defined by the act, in part, as "intended to affect the structure or any function in the body of man or other animals." Also, part of the FFDCA definition of "new animal drug" is one intended for use in animals that is not generally recognized as safe and effective for use under the conditions prescribed or recommended, and that has not been used to a material extent or for a material time.

"The rDNA construct in a GE animal that is intended to affect the structure or function of the body of the GE animal, regardless of the intended use of products that may be produced by the GE animal, meets the FFDCA drug definition," the guidance states (on page 5). A new animal drug is considered "unsafe"unless FDA has approved an application for that particular use, or it is for investigational use and subject to an exemption from the drug approval requirement (among a few other specified exemptions). Therefore, most developers will have to submit to the "Investigational New Animal Drug" (INAD) process at FDA prior to shipping any GE animals or to marketing any food or feed derived from GE animals. In other words, it is illegal to introduce food from a GE animal into the food supply that has not been approved by FDA. The guidance lays out the pre-market approval process including the information to be required of developers.

Under the guidance, FDA will examine both the direct toxicity (including allergenicity) potential of food from a GE animal as well as any indirect toxicity. Generally, food and feed will be considered safe if the composition of edible materials from the GE animal can be shown to be as safe as from a non-GE animal. The labeling requirements for GE-derived foods would be the same as for other foods: FDA has oversight over the labeling of seafood, dairy products, and whole shell eggs, and USDA over the labeling of most meat, poultry, and egg products (see below). More specifically, food from GE animals would not have to be so labeled, except when it takes on a different character from its non-GE counterpart.

U.S. Department of Agriculture (USDA)

Several USDA agencies, operating under a number of statutory authorities, also have at least potential roles in the regulation of transgenic and cloned animals and their products. As several critical reviews have indicated, USDA has not had a clearly spelled out policy in this area, including whether it intends to exercise these authorities to regulate GE animals.[15] USDA's Animal and Plant Health Inspection Service (APHIS) earlier had expressed its intention to publish an advance notice of proposed rulemaking (ANPR) on GE animals, possibly in 2008.[16] Instead, in concert with the FDA's notice on its draft guidance, APHIS published, in the September 19, 2008, *Federal Register*, a request for information from the public and scientists on how GE animals might affect U.S. animal health. Comments were accepted until November 18, 2008, as they had been for the FDA draft guidance. Further APHIS action was pending in mid- January 2009.

APHIS has broad authority, under the Animal Health Protection Act (AHPA; 7 U.S.C. §8301 *et seq.*) to regulate animals and their movement to control the spread of diseases and pests to farm- raised animals. APHIS also administers the Viruses, Serums, Toxins, Antitoxins, and Analogous Products Act (21 U.S.C. § 151-159), aimed at assuring the safety and effectiveness of animal vaccines and other biological products, including those of GM origin, and the Animal Welfare Act (7 U.S.C. §2131 *et seq.*), portions of which govern the humane treatment of several kinds of warm-blooded animals used in research (but generally not agricultural animals). Elsewhere at USDA, the Food Safety and Inspection Service (FSIS) is responsible for ensuring the safety and proper labeling of most food animals and meat and related products derived from them under the Federal Meat Inspection Act (21 U.S.C. §601 *et seq.*) and Poultry Products Inspection Act (21 U.S.C. §451 *et seq.*).

Other Authorities

Reports and studies have cited a number of other authorities and federal agencies that are or could be relevant. The National Environmental Policy Act (NEPA; 42 U.S.C. §4321 *et seq.*) requires federal agencies to consider the environmental impacts of their actions.[17] The Environmental Protection Agency derives its authority from, among other laws, the Federal Insecticide, Fungicide and Rodenticide Act (FIFRA; 21 U.S.C. §301 *et seq.*); pesticides derived from living organisms, including those of biotechnology, are within its purview. The Interior Department's Fish and Wildlife Service and the Commerce Department's National Marine Fisheries Service have also been cited.[18]

Cloning Policy Developments

In 1997, scientists at the Roslyn Institute in Scotland used nuclei from the mammary cells of an adult sheep to clone "Dolly." Such nuclear transfer (NT) techniques were first developed in amphibians in the 1950s. They were first used in sheep in 1986, with the production of clones using nuclei taken from sheep embryos. The significance of Dolly was that she was cloned from differentiated cell types obtained from an adult (called "somatic cell nuclear transfer" or SCNT), rather than undifferentiated cells from an embryo ("embryonic NT").

Cloning in animal agriculture is generally not applied in isolation from other biotechnologies such as genetic engineering. Scientists note that cloning does not require fertilization and is not, by itself, a form of genetic engineering, i.e., altering, removing, or inserting genes into an animal's existing DNA. However, cloning can involve transgenic as well as non-transgenic cells.

SCNT is not yet a notably efficient technique ready for widespread commercial adoption. For example, only about 6% of the embryos transferred to recipient cows resulted in healthy, longterm surviving clones, according to a 2005 report.[19] The European Food Safety Authority (EFSA) recently reported that overall success rates vary by species, ranging from 0.5% to 5%.[20]

Success rates are said to be improving, however. As more efficient cloning technologies, which can overcome the range of cloning abnormalities that have resulted from SCNT, are introduced, they could provide new opportunities in human medicine, agriculture, and animal welfare. This is the focus of much of the current international animal biotechnology research.

The EFSA draft scientific opinion estimated the number of live clones worldwide in 2007 to be less than 4,000 cattle and 1,500 pigs, of which about 750 cattle and 10 pig clones were in the United States. EFSA reported that life span data were limited, with only a few reports on cattle of six to seven years of age and no data available in 2007 on the full natural life span of livestock clones generally.[21]

FDA Risk Assessment

The FDA in July 2001 began asking companies to refrain voluntarily from marketing the products of cloned animals and their offspring until it could fully assess the scientific information on their safety. It subsequently issued two draft risk assessments on the matter, culminating in the release of a final risk assessment—and industry guidance that effectively lifted its voluntary moratorium—on January 15, 2008.

WHAT IS CLONING?

Cloning, or somatic cell nuclear transfer, is a process by which animals are reproduced asexually.... In cloning, a differentiated somatic cell (a non-germ line cell from an existing animal) is introduced to an oöcyte (a cell that is the immediate precursor of a mature egg) that has had its nucleus (and thus its genome) removed, and then, following some manipulations, is induced to start replicating. If all goes well, the dividing cell is implanted into a female animal (dam), continues to develop normally, and is delivered just as any newborn.

—Animal Cloning: A Risk Assessment, p.20

WHAT'S THE DIFFERENCE BETWEEN A CLONED AND A GE ANIMAL?

Clones covered by FDA's January 2008 risk assessment and guidance are "just clones"—that is, they are copies of individual conventionally-bred animals, and do not contain any rDNA constructs. What can be confusing is that an animal clone can be genetically engineered (i.e., have an rDNA construct introduced into it), and a GE animal can be reproduced by cloning. The September 2008 guidance covers GE animals, regardless of whether they were reproduced by cloning, but does not cover animal clones that do not contain an rDNA construct.

—Q and A to Accompany FDA's September 18, 2008 Draft Guidance

Some stakeholders, including several Members of Congress, wanted a continuation of the moratorium until more studies are completed on safety and other aspects of animal cloning. Their views were reflected in nonbinding language to accompany the omnibus spending measure for FY2008, which passed in late 2007, and binding language in the Senate version of the pending farm bill (H.R. 2419); see "Congressional Action," at the end of this report, for details.

In an October 2003 draft risk assessment, the FDA had concluded that "the current weight of evidence suggests that there are no biological reasons ... to indicate that consumption of edible products from clones of cattle, pigs, sheep or goats poses a greater risk than consumption of those products from their non-clone counterparts."[22] However, shortly after the assessment's publication, many members of the FDA's Veterinary Medicine Advisory Committee stated that there were not enough data to fully understand any potential risks from this relatively new technology.[23]

On December 28, 2006, the FDA's CVM released another long-awaited risk assessment—also in draft form, along with a proposed risk management plan and draft guidance for industry—on the safety of animal clones and their offspring.[24] The lengthy risk assessment examined two important questions: the safety of food from cloned animals and their progeny, and the effects of the process on the health of these animals. FDA officials stressed that the risk assessment did not address any other issues, such as the social and ethical aspects of cloning or consumer acceptance of cloned animal products. The risk assessment also focused only on cloning from non-transgenic cells. This risk assessment is now final.[25]

The final risk assessment concluded—as had the December 26, 2006 draft—that the meat and milk of clones of adult cattle, pigs, and goats, and the meat and milk from the offspring of these clones, were as safe for human consumption as the food from conventionally bred animals. The FDA added that not enough data were available to reach the same conclusions about sheep clones (or other species) and recommended that they not yet be used for human food. The risk assessment arrived at these conclusions after analyzing physiological, anatomical, health, and behavioral data on the animals and evaluating available information on the chemical composition of their milk and meat. FDA said its final assessment was peer-reviewed by an independent panel of experts, who agreed with the findings. The final assessment also took into account many thousands of public comments and additional data that became available since the draft was prepared, the agency said.

The agency said it would not require any special measures (including labeling) relating to the use of food products or animal feed derived from cloned cattle, goats, and pigs because they are "no different from food derived from conventionally bred animals. Should a producer express a desire for voluntary labeling (e.g., 'this product is clone-free'), it will be considered on a case-by-case basis to ensure compliance with statutory requirements that labeling be truthful and not misleading."[26] The industry guidance states that products from the offspring of cloned animals of any species are suitable for food or feed consumption.

Upon release of the 2006 proposed assessment, FDA officials said they were continuing to ask livestock breeders and producers to keep food products from cloned animals and their offspring out of commerce. This moratorium was announced in July 2001 and remained in effect until the final guidance was released in January 2008. Although the FDA moratorium is no longer in effect, USDA has since been encouraging technology providers to maintain the voluntary moratorium, but for products from cloned animals only. Food products from cloned animals' offspring would not be similarly constrained. During a continuation of this moratorium, USDA said it would "work closely with stakeholders to ensure a smooth and seamless transition into the marketplace for these products."[27]

Officials emphasized that, at least initially, almost all cloning-related foods will come not from the clones themselves but rather from their sexually reproduced offspring. Cloned animals—like other "elite" breeding animals such as a prized bull—are too valuable to use for food production, except possibly when they reach the end of their productive lives. The FDA observed that it can cost $20,000 or more to produce one such animal.

One ongoing question has been whether, in fact, milk and meat from the offspring of clones are entering the food supply. Several press accounts have quoted farmers and others who say they have sold such offspring to be slaughtered for food, and that the total number nationwide may be in the hundreds or even thousands. On the other hand, those numbers would amount to a small fraction of the total number of U.S. livestock slaughtered (e.g., 34 million cattle and 109 million hogs in 2007).[28]

FDA's January 2008 risk assessment did conclude that some animals involved in cloning (notably cattle and sheep surrogate dams) and some clones are at greater health risk than conventional animals. While the types of animal health problems observed in cloned animals are no different than those found with other assisted reproductive technologies, these problems appear more frequently in cloning than in the other technologies. Such problems include late gestational complications in the surrogate mothers, and increased risk of mortality and morbidity in calf and lamb clones that are apparently caused mainly by large offspring syndrome (LOS). Swine and goat clones and their mothers do not appear to experience any additional cloning-related problems, FDA reported.

FDA found that livestock clones as a group tend to have more health problems and death rates at or right after birth. However, the risk assessment added that most animals surviving the neonatal period appear to grow and develop normally, and that no increased risk of adverse health effects have been reported in clones approaching reproductive maturity. The agency said that the technology was too new to draw any conclusions on the relative longevity of livestock clones.

European Views

Cloning currently is not in commercial use in Europe, nor is there any specific authorization procedure in the European Union for products from cloned animals.[29] The European Food Safety Authority (EFSA) on July 24, 2008, released its scientific opinion on the food safety, animal health and welfare, and environmental impacts of animals derived from cloning. The EFSA opinion was limited to cattle and pigs, but its findings generally do not appear to be substantially different from those in the FDA assessment. According to EFSA, "Based on current knowledge, and considering the fact that the primary DNA sequence is unchanged in clones, there is no indication that differences exist in terms of food safety between food products from healthy cattle and pig clones and their progeny, compared with those from healthy conventionally-bred animals." As with FDA, the EFSA conclusion is based on an examination of compositional and nutritional data, the probability for the presence of novel constituents, the health status of the animal, and available data on toxicity, allergenicity, and microbiology.

EFSA agreed that SCNT can be a successful reproductive technique, although death and disease rates of clones are significantly higher than those of conventionally reproduced animals. EFSA noted that surrogate dams have higher rates of failed pregnancies and other problems. "A significant proportion of clones, mainly within the juvenile period for bovines and perinatal period for pigs, has been found to be adversely affected, often severely and with fatal outcome. Most clones that survive the perinatal period are normal and healthy," however. Adverse health effects were not observed in the offspring of clones, although studies have not been conducted for their entire lifespans.[30]

EFSA also found no indication of new or additional environmental risks compared with conventionally bred animals.

The EFSA report findings are being used by the European Commission (EC), the European Parliament, and member states as they discuss their policies regarding animal cloning. There is disagreement over the issue among officials in the various European bodies. This was illustrated when the European Parliament in early September 2008 approved a resolution asking the EC to propose a comprehensive ban on cloning animals for food, out of animal health and welfare and ethical concerns.[31] However, at an initial discussion of the issue in early January 2009, EC commissioners indicated that there was time to consider a science-based risk assessment before deciding whether or not to ban cloning.[32]

Other Views

Critics, including a number of consumer advocacy and animal rights groups, oppose the use of cloned animals and their offspring for food. In October 2006, a coalition formally petitioned the FDA to impose a moratorium on producing foods from cloned animals, and to establish rules for mandatory pre-market review and approval of cloned foods by regulating clones as new drugs under the food and drug act. These consumer and animal welfare groups contend that FDA has ignored or minimized a number of food safety and animal welfare problems. These include the treatment of surrogate mothers with high doses of hormones and their clone offspring, who often have severely compromised immune systems, with large doses of antibiotics, which could enter the human food supply; imbalances in clones' hormone, protein, or fat levels that could compromise meat and milk safety and quality; the possibility of increased foodborne illnesses; and a wide variety of health problems and abnormalities in the animals themselves. Consumer advocates assert that there are no consumer benefits from cloning.

FDA disagrees, citing the potential to breed livestock that meet consumers' changing tastes for traits like leanness, tenderness, color, size of meat cuts, and so forth. These would be in addition to potential producer-related benefits such as disease resistance, climate suitability, fertility, and improved physical qualities, FDA states.[33]

Opinion polls have suggested that the general public may not yet be ready for widespread animal cloning. A 2006 poll commissioned by the Pew Initiative on Food and Biotechnology, for example, found that Americans generally were not well informed about animal cloning, but 64% of those questioned are "uncomfortable" with it. Forty-three percent of Americans believed that foods from clones are unsafe, the survey found. Cloning opponents cite these types of findings to argue that products from cloned animals should be labeled so that consumers could avoid them if they wanted to. Some critics have urged the President to halt all FDA actions on cloning until a national panel that includes ethicists and religious leaders can consider the ethical and other social issues that they believe the technology has raised.

The NRC animal biotechnology report had stated that embryonic splitting and nuclear transfer using embryonic (not adult) cells were performed with some dairy cows to successfully produce genetically valuable offspring that were milked commercially and whose milk and meat did enter the food supply. Few concerns were raised by NRC authors about using these types of cloned animals for food, since they are generally believed to pose a low level of food safety concern. However, evaluating cloned-animal food composition "would be prudent to minimize any food safety concerns. The products of offspring of cloned animals were regarded as posing no food safety concern because they are the result of natural matings."[34] Other issues, notably consumer acceptance, social values, and animal welfare, could eventually overshadow any lingering questions about human health.

OTHER POLICY CONCERNS

The following are among the policy concerns that have arisen along with the development of new biotechnologies in animal agriculture. Some may be more applicable to GE-related technologies than to cloning *per se*, although others, like social acceptance and animal welfare concerns, may apply to both.

Environmental Issues

Environmental concerns arising from emerging animal biotechnologies are largely speculative at this time because few products have been commercialized. For example, although the EFSA draft scientific opinion foresaw no environmental impact, it also noted that limited data were available on this aspect of animal cloning.

Industrial developers of agricultural biotechnology might argue that more efficient production of animal-based feeds could reduce the resources necessary to produce food and, thereby, reduce the environmental burden of animal production. Should the development and widespread adoption of the "EnviroPig" (tm), which produces less phosphorus in its waste, occur, it might be considered by some to be a positive environmental benefit of agricultural biotechnology.

The 2002 NRC animal biotechnology report noted potential negative environmental impacts of genetically altered animals. Escape, survival, and gene flow into wild populations were identified as major concerns. Of most concern to the NRC committee was the escape into the environment of "super" salmon that have been genetically modified for rapid growth, and the likelihood that they could then breed with wild populations in the environment.[35] Other genetically altered animals such as fish, insects, and shellfish could also potentially escape into natural environments and become feral, disrupt ecosystems, or introduce novel genes in a natural population.

The FDA guidance on GE animals notes that the agency will comply with requirements of the National Environmental Policy Act (NEPA). Environmental risks are likely to differ depending upon the animal and application. For example, the types of environmental concerns arising from a GE cow bred for resistance to mastitis will differ greatly from the concerns raised by a GE freshwater fish engineered to grow more rapidly.[36] Material to accompany the FDA's final guidance notes that "although the agency has extensive experience in environmental assessment, including for fish, other federal and state agencies have overlapping or complementary authority and expertise" that FDA intends to tap. It also promised to make the results of environmental reviews public.[37]

Food Safety

Unexpected and unintended compositional changes arise with all forms of plant and animal genetic modification, including GE, concluded the IOM-NRC report on genetically engineered foods.[38] The report added that, so far, no GE-related adverse human health effects have been documented. However, the report's authors cited "sizeable gaps" in the ability to identify compositional changes caused by all forms of genetic modification—whether GE or conventional—and their relevance for human health, and they recommended new approaches for assessing the safety of new foods both before and after they enter the market.

Previous research and experience with commercializing transgenic plants suggested that negative effects on human health were virtually nonexistent. While not asserting that genetically modified organisms necessarily generate health problems, more recently reported research in peer-reviewed scientific journals has suggested that GMOs may raise food safety concerns:

- Australian researchers have published an article explaining that the transfer from a bean to a pea gene that expresses an insecticide protein has resulted in antibody production in mice fed the transgenic pea. The antibody reaction is a marker of allergic reaction.[39]
- Italian researchers at the University of Urbino had previously shown that absorption of transgenic soy by mice induced modifications in the nuclei of their liver cells. Recent research showed that a return to non-transgenic soy made the observed differences disappear.[40]
- Norwegian scientists at the University of Tromso demonstrated that the catalyst 35S CaMV, an element of the genetic structures used to modify a plant, can provoke gene expression in cultured human cells. This catalyst was previously believed to operate in this way only in plants.[41]

In the NRC animal biotechnology report, experts observed that the scientific principles for assessing the safety of GE animals are "qualitatively the same" as for non-GE animals.[42] However, because GE can introduce new proteins into foods, the potential for allergenicity, bioactivity, and/or toxicity responses should be considered, they said. Others have remarked that animals genetically engineered for nonfood products like pharmaceuticals or replacement organs might be of concern if such animals entered *or* affected the food supply.

Consumer and Social Acceptance

Criteria for selecting desirable traits to be produced through transgenic animals will likely be based on the demand for specific commercial characteristics. Even if scientific evidence is convincing that GE and cloned animal products are safe and beneficial for human consumption or economically valuable to producers, other concerns may limit marketplace and consumer acceptance.

Polls in recent years in the United States indicate that public knowledge about food and biotechnology generally remains limited. In two 2005 surveys, approximately half of those surveyed expressed opposition to the use of biotechnology in the food supply.[43] More than half of those in a 2005 Pew-sponsored poll said they opposed research into genetically modified animals, although opposition declined with increased knowledge. Many Americans have heard about animal cloning; two-thirds expressed discomfort with it—more of them out of religious or ethical concerns than food safety concerns. A majority of respondents to the Pew survey believe that regulators should take into account ethical and moral considerations. (See also the 2006 Pew findings on cloning, under "Cloning Policy Developments," section on "Other Views.")

Consumers may be less willing to accept the practice of genetically modifying animals than plants, some have argued, observing that people relate differently to animals, which many recognize as sentient beings. Some observers have expressed the concern that cloning farm animals might lead more quickly to human and pet cloning, which those observers oppose. Others believe that modifying animals, for example, to save human lives through xenotransplantation or the production of some important drug, might be more acceptable than doing so simply to produce more or cheaper food.

Further, science alone cannot resolve ethical views that appear to vary widely:

> Some people, irrespective of the application of technology, consider genetic engineering of animals fundamentally unethical. Others, however, hold that the ethical significance of animal biotechnologies must derive from the risk and benefits to people, the animals, and/or the environment. Yet another view focuses on the right of humans to know what they are eating or how their food or pharmaceuticals are being produced and therefore labeling becomes an issue to be addressed.[44]

Food industry leaders appear sensitive to consumer unease with animal biotechnology in general and cloning in particular. Many companies are not yet ready to process and sell meat and milk from transgenic and cloned animals because, they believe, consumers may view the products as less safe than more conventionally produced food products. The food industry also does not want to be portrayed as overstepping any widely held ethical or moral concerns about the new technologies.[45]

Such observations led some skeptics of animal biotechnology to propose that the FDA not only consider the science and safety issues, but also these broader concerns. In the area of human reproductive health, for example, the FDA and other federal agencies have invoked particular moral arguments either to reinforce scientific arguments or to counterbalance scientific evidence. Others believe that the FDA should base its decisions only on scientific evidence, and perhaps some other body should be established to consider the ethical and cultural questions. As the NRC animal biotechnology report observed, regulatory decisions and enforcement involving animal biotechnology "are difficult in the absence of an ethical framework."[46]

Labeling

Some believe that segregating and labeling the products of biotechnology in agriculture, including meat and milk, would enable the consumer to choose whether or not to buy such products. Either segregating and labeling biotechnology products or failing to do so could contribute to public suspicion that these products are flawed or different in some negative way, which may lead to contradictory policy decisions.[47] As noted, the FDA guidance on GE animals does not require their products to be so labeled. Opponents of labeling argue that biotechnology products essentially are the same as more conventional products—and are subjected to the same rigorous safety standards—and therefore should be treated no differently in the marketplace. A study by USDA's Economic Research Service reported that consumers' willingness to pay for a food item declines when the food label indicates that it was produced with the aid of biotechnology.[48]

In December 2007, two leading U.S. livestock cloning companies announced they were creating a voluntary system for tracking cloned animals. The program, which they asserted "is designed to facilitate marketing claims," is to involve a national registry providing for the individual identification of each cloned animal, affidavits that owners will sign committing to proper marketing and disposal of such animals, and monetary deposits to be returned when such commitments are fulfilled. Consumer organizations immediately criticized the initiative, noting that it would not track the offspring of cloned animals and was not mandatory.[49]

However, fierce debate continues regarding even the *voluntary* labeling of products to denote that they are *not* derived from biotechnology—notably the use of so-called "rBST-free" labels on dairy products. Although FDA cleared rBST as safe for commercial use in 1994, and it has been widely adopted by U.S. dairy farmers, the substance continues to be viewed skeptically by some consumer advocates. They have convinced a number of dairy processors and retailers to label their products as free of added BST manufactured through the use of recombinant DNA technology. FDA guidance permits such labeling so long as it is not misleading, is presented in the proper context, and is adequately substantiated. For example, a firm cannot state that its product is "BST-free" because the substance occurs naturally in milk, or even simply that it is "rBST-free" because that could imply a compositional difference between naturally occurring and rBST, which there is not, FDA has ruled. But certain claims that no rBST was used could be acceptable, the agency stated.[50]

FDA has looked to the states to evaluate the acceptability of such labels, and a number of these states have challenged processors' use of them. Battles over such restrictions have been notable in Pennsylvania, Ohio, Utah, Kansas, and elsewhere.[51]

Meanwhile, USDA has reminded stakeholders that products from cloned animals are not eligible to be labeled as organic, under its National Organic Program (NOP). While cautioning that the organics program is a marketing, not a food safety, program, USDA also noted: "Cloning as a production method is incompatible with the Organic Foods Production Act and is prohibited under the NOP regulations."[52] However, the status of products from clone offspring is less clear. The Department noted that USDA's Agricultural Marketing Service, where the NOP is located, was preparing rulemaking to address the organic status of such offspring.[53]

Animal Welfare

Some aspects of gene transfer, and of cloning, have the potential to create infectious disease hazards and/or impaired reproduction. Looming large in the ethical debate are questions about whether genetic modifications, cloning, and other technologies stress animals unnecessarily, subject them to higher rates of disease and injury, and hasten death.[54] The NRC animal agriculture report noted, for example, that ruminants produced by *in vitro* culture or nuclear cell transfer methods tend to have higher birth weights and longer gestation periods than those produced by artificial insemination, creating potential calving problems. Nuclear transfer techniques to propagate genetic modifications may increase risks to the reproductive health and welfare of both the surrogate female animals and their transgenic offspring. The report cited other evidence of problems such as anatomical, physiological, or behavioral abnormalities in many transgenic animals.[55] Some scientists have countered that animal welfare problems have been exaggerated and tend to recede, particularly as the technologies are perfected. Most appear to agree, however, that animals originating from some forms of genetic modification or from cloning may require closer observation and care. (See also the section of this report on Cloning Policy Developments.)

In Europe, officials continue to consider the ethical dimensions of animal welfare in formulating their own cloning policies. The EC charged a separate panel, the European Group on Ethics in Science and New Technologies, with drafting an opinion. This group released an opinion on January 16, 2008, stating, in part:

> Considering the current level of suffering and health problems of surrogate dams and animal clones, the EGE has doubts as to whether cloning animals for food supply is ethically justified. Whether this applies also to progeny is open to further scientific research. At present, the EGE does not see convincing arguments to justify the production of food from clones and their offspring.

If such products were to be allowed into the market, certain requirements regarding food safety, animal welfare and traceability should be met, the group added.[56]

Genetic Diversity

Could the introduction of a few genetically altered or cloned "superspecies" bring too much genetic uniformity to herds? As genetic diversity declines, herds could be more susceptible to diseases, leading to large production losses and/or much heavier use of antibiotics and other animal drugs to treat them, some have argued. A related concern is that a relative handful of "elite" producers or breeders might hold the proprietary rights to these species, to the disadvantage of many farmers and ranchers. Some animal biotechnology researchers have pointed to the potential importance of preserving unaltered germlines in domestic animals because they could prove to be an invaluable "gene bank" in the event that novel infectious diseases or inheritable genetic defects were inadvertently introduced into modified subpopulations as a consequence of genetic modification.

Trade Issues

If the United States were to be the first country to approve food products from cloned animals, how might the decision affect U.S. exports? Any exports of the products of animal biotechnology would presumably encounter a wide spectrum of foreign regulatory regimes, some more restrictive than the U.S. system. For example, the current European Union restriction on new biotechnology products is likely to encompass various restrictions on animal biotechnology as it does on plant biotechnology. On the other hand, researchers in a number of other countries, including some EU members, have been producing clones, and one—France—has published its own risk assessment on clones that, FDA has observed, generally agrees with the U.S. assessment.[57]

International guidelines pertaining to exports of animal products derived from biotechnology are being considered. The Codex Ad Hoc Intergovernmental Task Force on Foods Derived from Biotechnology held an initial meeting in September 2005 in Chiba, Japan, to determine the new work projects. Almost every country, except for the United States, proposed an animal biotechnology project. The task force agreed to move forward with a recombinant DNA (r-DNA) animal project, specifically to develop guidelines for how countries would assess the safety of foods derived from r-DNA animals.[58] At a meeting in November-December 2006 in Chiba, the task force, among other things, reviewed the developing guidelines, agreeing to limit their scope to food safety and nutritional issues (while recognizing the importance of others like animal welfare, environmental, and ethical concerns).[59]

USDA has said it would consider "discussion with industry on possible verification of its supply chain management plan to ensure that trading partners are aware of whether or not they receive cloned or non-cloned products."[60]

Imports of GE animals and their products into the United States also are of concern to some. In a December 2008 audit report, USDA's Office of Inspector General (OIG) concluded that the Department has not established an import control policy for such imports but needs to do so. "To mitigate any risks to the U.S. environment, agriculture, and commerce from unapproved transgenic plants and animals entering the U.S. food supply, USDA will need to monitor" foreign developments in such technologies more closely, OIG also noted.[61]

CONGRESSIONAL ACTIVITY

Members of Congress in the past have proposed various bills aimed at more closely regulating the plant and animal products of agricultural biotechnology generally, requiring such products to be labeled, and providing a means to recover any damages caused by the technology. As with human cloning, ethical issues concerning animal clones and other animal biotechnologies also may continue to be visible public issues. As before, the 111th Congress could be asked to play a larger role in weighing the benefits and costs of these evolving technologies, and to refine existing government oversight.

In the 110th Congress, the Senate-passed version of the 2007-2008 farm bill (H.R. 2419, in Section 7507) contained statutory language that would have required the FDA both to postpone publication of its final risk assessment until the completion of several newly mandated studies, and also to maintain the voluntary marketing moratorium until then. However, conferees deleted the Senate language from the final measure (signed into law as P.L. 110-234).

More specifically, the Senate bill would have directed the HHS Secretary to contract with the National Academy of Sciences to conduct a study on the safety of food products from cloned animals and the health effects and costs attributable to milk from cloned animals, with a report to Congress due within one year of enactment. The study was to address whether there were a sufficient number of studies to support the FDA draft assessment, and whether there were other pertinent ones that were not taken into account. It would have included an evaluation of potential public health effects and associated health care costs, and an evaluation of any consumer behavior and negative health and nutrition impacts resulting from a decrease in dairy consumption if milk from cloned animals and their offspring is commercialized.

Animal cloning-related bills (S. 414 and H.R. 992) were introduced in early 2007, both of which would have amended federal food safety laws to require that foods from cloned animals and their offspring be so labeled. Also introduced in early 2007 was S. 536/H.R. 1396, which would have prohibited the use of the "organic" label on food products from cloned livestock or their offspring. Another (H.R. 4855), introduced in late 2007, would have required studies like those in the Senate farm bill on the impacts of food products from cloned animals entering the food supply.

In late July 2008, three related bills were introduced that, their sponsor said, were to create a comprehensive framework for regulating GMOs. H.R. 6636 would have required the labeling of all foods produced with GE material. H.R. 6635 would have prescribed relatively stringent new regulations for FDA approval and oversight of GE crops. H.R. 6637 would have regulated business dealings between agricultural producers and the developers of genetically engineered plants and animals, and sought to hold biotechnology companies liable for any adverse on-farm impacts from their products.

Although these bills were not enacted, they could re-emerge, along with others, in the 111th Congress.

End Notes

[1] CVM, FDA, Guidance for Industry: *Regulation of Genetically Engineered Animals Containing Heritable rDNA Constructs*. The guidance can be accessed through the FDA website at http://www.fda.gov/OHRMS/DOCKETS/98fr/ FDA-2008-D-0394-gdl.pdf.

[2] FDA documents were posted on the Internet at http://www.fda.gov/cvm/cloning on January 15, 2008.

[3] National Research Council (NRC), *Animal Biotechnology: Science-Based Concerns*, (Washington, DC, National Academy Press, 2002). (Hereafter cited as NRC, *Animal Biotechnology.*) Unless noted, this CRS report is based on material in that document and the following additional sources: Institute of Medicine (IOM) and NRC, *Safety of Genetically Engineered Foods: Approaches to Assessing Unintended Health Effects* (Washington, DC, National Academy Press, 2004). (Hereafter cited as IOM-NRC, *Safety of Genetically Engineered Food.*); Council for Agricultural Science and Technology (CAST), *Biotechnology in Animal Agriculture: An Overview* (Issue Paper 23) (Washington, DC, February 2003). (Hereafter cited as CAST, *Animal Agriculture.*); Pew Initiative on Food and Biotechnology, *Issues in the Regulation of Genetically Engineered Plants and Animals*, (Washington, DC, Pew Initiative April 2004. See http://pewagbiotech.org/research/regulation (Hereafter cited as Pew, *Issues in Regulation.*); Pew Initiative on Food and Biotechnology, *Post-Market Oversight of Biotech Foods* (Washington, DC, Pew Initiative April 2003) (and other Pew materials); information accessed through the websites of the Biotechnology Industry Organization at http://www.bio.org/; the Center for Food Safety, at http://www.centerforfoodsafety.org/; the Union of Concerned Scientists, at http://www.ucsusa.org/; and the Consumer Federation of America, at http://www.consumerfed.org/.

[4] *New York Times*, March 27, 2006.

[5] FDA, "A Primer on Cloning and Its Use in Livestock Operations," at http://www.fda.gov/cvm/CloningRA_Primer.htm.

[6] *Genetic engineering* (GE) here refers to the use of molecular biology to alter cells by inserting or removing genes. GE is a form of *genetic modification (GM)*, which refers more broadly to the practices of altering an organism's genetic composition by both GE and non-GE methods.

[7] See CRS Report RL3 1358, *Human Cloning*, by Judith A. Johnson and Erin D. Williams.

[8] National Institutes of Health, National Human Genome Research Institute, "Knockout Mice," accessed January 21, 2009, at http://www.genome

[9] NRC, *Animal Biotechnology*, p. 14.

[10] See Center for Food Safety website, at http://www.centerforfoodsafety.org/geneticall7.cfm; and the Union of Concerned Scientists website, at http://www.ucsusa.org/food. Also see discussion of FDA's guidance for industry, later in this section.

[11] The FDA, as of early 2009, has not granted approval for any human foods from transgenic (or cloned) animals, although a "very limited number have been approved for rendering into animal feed components." The only FDA- approved product of biotechnology in wide commercial use is bST, and an application is pending for a GE salmon. (*Questions and Answers about Transgenic Fish*, at http://www.fda.gov/cvm/transgen.htm).

[12] Also see CRS Report RL34247, *Federal Regulation of Substances Generally Recognized As Safe (GRAS) and the Use of Carbon Monoxide in Packaging for Meat and Fish*, by Vanessa K. Burrows and Cynthia Brougher.

[13] NRC, *Animal Biotechnology*, p. 163. At a January 9, 2009 meeting, the FDA Blood Products Advisory Committee ruled that a new drug to prevent clots in humans, recombinant human antithrombin III produced in the milk of GE goats, is safe and effective. The FDA could make a final regulatory decision to approve the license application of its developer, GTC Biotherapeutics, soon. Source: "FDA issues final guidance to industry on transgenic animals," *Food Chemical News*, January 19, 2009.

[14] *Guidance for Industry: Regulation of Genetically Engineered Animals Containing Heritable rDNA Constructs.* FDA noted that much of the new guidance will be relevant also to non-heritable rDNA constructs (such as modifications intended for gene therapy); a separate guidance for non-heritable constructs might come later. The agency states at the outset: "This guidance represents the Food and Drug Administration's (FDA's) current thinking on this topic. It does not create or confer any rights for or on any person and does not operate to bind FDA or the public. You can use an alternative approach if the approach satisfies the requirements of the applicable statutes and regulations."

[15] See, for example, Pew, *Issues in Regulation.* Beginning on p. 139, the report contains an extensive discussion on how these and several other USDA authorities might be utilized for oversight of animal biotechnology.

[16] APHIS in 2007 established an Animals Branch within its Biotechnology Regulatory Services "to develop a regulatory framework for the possible regulation of genetically engineered animals." Source: APHIS website, "Regulation of Genetically Engineered Animals," at http://www.aphis.usda.gov/biotechnology news_transgenic_animals.shtml.

[17] The FDA guidance discusses how NEPA requirements will apply to the GE animal approval process.

[18] The FDA guidance on GE animals states that the agency will work with other relevant federal and state agencies should it receive a request for approval of a GE animal intended for release into the wild.

[19] D. N. Wells, "Animal Cloning: Problems and Prospects," *Review of Science and Technology*, 24(1): 251-264, 2005.

[20] European Food Safety Authority. *Draft Scientific Opinion on Food Safety, Animal Health and Welfare and Environmental Impact of Animals Derived from Cloning by Somatic Cell Nucleus Transfer (SCNT) and their Offspring and Products Obtained from the Animals*, Enorsed for public consultation December 19, 2007 and released January 11, 2008. Hereinafter cited as EFSA draft scientific opinion. Accessed on the Internet at http://www.efsa.europa.eu/EFSA/ efsa_locale-1178620753812_1178676922939.htm.

[21] Ibid.

[22] See http://www.fda.gov/cvm/Documents/CLRAES.pdf.

[23] "Panel Calls for More Data on Animal Cloning Risks," *Food Chemical News*, November 10, 2003.

[24] Availability of the draft documents was formally published in the January 3, 2007, *Federal Register* (72 FR 135-137; FDA docket no. 2003N-0573); comments were accepted until May 3, 2007.

[25] FDA documents were posted on the Internet at http://www.fda.gov/cvm/cloning on January 15, 2008. They are the Risk Assessment, Risk Management Plan, and Guidance for Industry.

[26] "FDA Issues Documents on the Safety of Food from Animal Clones," FDA press release, January 15, 2008.

[27] Statement by Bruce Knight, Under Secretary for Marketing and Regulatory Programs on FDA Risk Assessment on Animal Clones, January 15, 2008.

[28] See, for example, "Animal Clones' Offspring Are in Food Supply," *The Wall Street Journal* online, September 2, 2008.

[29] "EFSA launches its draft opinion on animal cloning for public consultation," EFSA press release, January 11, 2008.

[30] "Scientific Opinion of the Scientific Committee on a request from the European Commission on Food Safety, Animal Health and Welfare and Environmental Impact of Animals derived from Cloning by Somatic Cell Nucleus Transfer (SCNT) and their Offspring and Products Obtained from those Animals," *The EFSA Journal* (2008) 767, 1-49; http://www.efsa.europa.eu/EFSA/efsa_locale-1178620753812_1211902019540.htm.

31 "European Parliament urges ban on cloning for food," *Food Chemical News*, September 8, 2008.

32 "EU not ready to ban animal cloning for food," *Food Chemical News*, January 19, 2009.

33 "A Primer on Cloning and Its Use in Livestock Operations," at http://www.fda.gov/cvm/CloningRA_Primer.htm.

34 CAST, *Animal Agriculture*, p. 8.

35 NRC, *Animal Biotechnology*, p. 73.

36 FDA, *Consumers Q&A*, accompanying its September 18, 2008, draft guidance.

37 "FDA's Response to Public Comments," accessed January 21, 2008, at http://www.fda.gov/cvm/GEanimals.htm #Consumer%20Health%20Information.

38 IOM-NRC, *Safety of Genetically Engineered Food*, p. 1.

39 V. E. Prescott, P. M. Campbell, et al., "Transgenic Expression of Bean-Amylase Inhibitor in Peas Results in Altered Structure and Immunogenicity," *Journal of Agriculture and Food Chemistry*, 53(23); 9023-9030, November 2005.

40 M. Malatesta, C. Tiberi, et al., "Reversibility of Hepatocyte Nuclear Modifications in Mice Fed on Genetically Modified Soybean," *European Journal of Histochemistry*, 49(2):237-242, July-September 2005.

41 M. R. Myhre, K. Fenton, et al., "The 35S CaMV Plant Virus Promoter Is Active in Human Enterocyte-like Cells," *European Food Research and Technology*, 222(1-2):185-193, January 2006.

42 NRC, *Animal Biotechnology*, p. 65.

43 Gallup poll on biotechnology and food safety, July 2005; Mellman Group/Public Opinion Strategies Poll Conducted for the Pew Initiative on Food and Biotechnology, October 2005. See http://www.pewagbiotech.org (accessed March 20, 2006).

44 NRC, *Animal Biotechnology,* p. 13.

45 These concerns were explored in depth at an October 2006 symposium, "Animal Biotechnology: Considering Ethical Issues," sponsored by the Pew Initiative on Food and Biotechnology and by Michigan State University. Also reported by *Food Chemical News*, October 23, 2006.

46 Ibid.

47 C. T. Foreman, *Can U.S. Support for Food Biotechnology Be Salvaged?*, paper prepared for the American Enterprise Institute for Public Policy Research Conference on Biotechnology, the Media and Public Policy, June 12, 2003.

48 A. Tegene, W. Huffman, et al., *The Effects of Information on Consumer Demand for Biotech Foods: Evidence from Experimental Auctions*, USDA, Economic Research Service, Technical Bulletin 1903, March 2003.

49 Sources: "Top U.S. Cloning Companies Announce New System to Track Cloned, Livestock," December 19, 2007, Viagen press release, and "Consumer Groups Displeased with Voluntary Animal Clone Registry," *FDA Week*, December 21, 2007.

50 FDA, "Interim Guidance on the Voluntary Labeling of Milk and Milk Products From Cows That Have Not Been Treated With Recombinant Bovine Somatotropin," *Federal Register*, February 10, 1994.

51 See, for example, "Dairy processors fight to save 'rBST-free' labels, March 3, 2008, and "Kansas is latest battleground in 'rBST-free' labeling fight," December 22, 2008, both in *Food Chemical News*.

52 USDA, Questions and Answers, FDA's Final Risk Assessment, Management Plan and Industry Guidance on Animal Clones and their Progeny, January 2008.

53 Ibid. Such a rule was not unveiled in 2008.

54 See also H. P. S. Kochhar, G. Adlakha-Hutcheon, and B. R. Evans, "Regulatory Considerations for Biotechnology- Derived Animals in Canada," *Review of Science and Technology*, 24(1):1 17-125, 2005.

55 NRC, *Animal Biotechnology*, p. 11.

56 "Ethical aspects of animal cloning for food supply," Opinion No. 23. The group's website is at http://ec.europa.eu/ european_group_ethics/index_en.htm.

57 Stephen F. Sundlof, then Director of FDA Center for Veterinary Medicine, transcript of a December 28, 2006, teleconference on the draft risk assessment.

58 Codex is recognized by the World Trade Organization (WTO) as the body that sets food safety standards for facilitating international trade of food products. The WTO cites Codex texts as a benchmark in the Agreement on Sanitary and Phytosanitary Measures (SPS).

59 The collaborative document is entitled *Proposed Draft Guideline for the Conduct of Food Safety Assessment of Foods Derived from Recombinant-DNA Animals*. Codex adopted the guidelines at its Geneva meeting on June 30-July 5, 2008.

60 Questions and Answers, FDA's Final Risk Assessment, Management Plan and Industry Guidance on Animal Clones and their Progeny.

61 USDA, OIG, *United States Department of Agriculture Controls Over Importation of Transgenic Plants and Animals*, Audit Report No. 50601-17-Te, December 2008.

In: Transformation of U.S. Animal Agriculture
Editor: Justin M. Daigle

ISBN: 978-1-60876-938-4

Chapter 3

THE CHANGING ECONOMICS OF U.S. HOG PRODUCTION

Nigel Key and William McBride
U.S. Dept. of Agriculture (USDA)

ABSTRACT

The increasing size and specialization of hog operations reflect structural change in U.S. swine production during the past 15 years. The number of farms with hogs has declined by over 70 percent, as hog enterprises have grown larger. Large operations that specialize in a single phase of production have replaced farrow-to-finish operations that performed all phases of production. The use of production contracts has increased. Operations producing under contract are larger than independent operations and are more likely to specialize in a single phase of production. These structural changes have coincided with substantial gains in efficiency for hog farms and lower production costs. Most of these productivity gains are attributable to increases in the scale of production and technological innovation. Productivity gains likely contributed to a 30-percent reduction in the price of hogs at the farm gate.

Keywords: Hogs, farm productivity, production contracts, pork prices, scale of production, farm structure, total factor productivity

SUMMARY

The increasing size and specialization of hog operations reflect significant structural change in U.S. swine production during the past two decades. Once dominated by small operations that practiced crop and hog farming, the industry has become increasingly concentrated among large operations that produce hogs on several different sites. Further, large operations that specialize in a single phase of production have replaced farrow-to-finish operations that performed all phases of production. Organizational change in hog production, particularly the widespread use of contracting, has enabled individual producers to grow by specializing in a single phase of production. Technological innovation has also been a driving force behind the changes and has contributed to substantial increases in farm productivity.

What Is the Issue?

As the industry has changed, hog producers have had to adjust the size, organizational structure, and technological base of their operations, or cease production. The effects of the changes have extended beyond the industry, as restructuring has heightened environmental risks and nuisance impacts, raised concerns about the integrity of rural communities in farming-dependent areas, precipitated controversy over animal welfare, and lowered pork prices for consumers. By providing information about changing structural characteristics and economic relationships in hog production, and what these suggest for the future of hog farms, this report provides context for these broader issues as well.

What Did the Study Find?

Scale and organization - The number of hog farms fell by more than 70 percent between 1992 and 2004, whereas the hog inventory remained stable. The average hog operation grew from 945 head in 1992 to 2,589 head in 1998 and to 4,646 head in 2004. The share of the hog inventory on operations with 2,000 or more head increased from less than 30 percent to nearly 80 percent. Operations with 5,000 or more head held more than 50 percent of the hog inventory in 2004.

Traditional farrow-to-finish production has given way to operations specializing in a single phase of production. Specialized finishing operations increased their share of output from 22 to 77 percent during 1992-2004, whereas the share of production from farrow-to-finish operations fell from 65 to 18 percent. Hog operations organized under production contracts grew from 3 percent of operations in 1992 to 28 percent in 2004 and accounted for more than two-thirds of hog production (sales and removals) in 2004. Operations producing under contract were larger than independent operations and were more likely to specialize in a single phase of production.

Regional trends - The rapid growth of hog operations along the east coast of the United States during 1992-98 slowed in subsequent years partly because the North Carolina State legislature placed a moratorium on expanded hog production in the State (a leading hog producer) in response to environmental concerns. In contrast, the size of hog operations increased more rapidly in Midwestern hog-producting States during 1998-2004 as contract production expanded in those areas.

Productivity gains- Structural change in the industry coincided with substantial gains in efficiency for hog farms, particularly on specialized hog-finishing operations. Feeder-to-finish operations had annual reductions in the amount of feed and labor used per unit of output of 4.7 percent and 13.8 percent, respectively, between 1992 and 2004, while their real, or inflation-adjusted, production costs per hundredweight of gain declined at an average annual rate of 4.7 percent.

For feeder-to-finish farms, total factor productivity increased at an average annual rate of 6.4 percent from 1992 to 1998 and 6.3 percent from 1998 to 2004. Most of these productivity gains were attributable to increases in the scale of production (scale efficiency) and technological innovation. Increases in the size of production operations helped account for almost half of the total increase in farm productivity. Further increases in scale efficiency likely will be limited for large farms. However, there is greater scope for efficiency gains in the sector as a whole from further increases in scale.

Trends in farm productivity in two major hog-producing regions, the Southeast and the Heartland, mirrored trends in farm output: productivity increased more in the Southeast between 1992 and 1998 and increased more in the Heartland between 1998 and 2004. Growth in average farm size and the resulting improvements in scale efficiency accounted for most of the differences in productivity growth between the Heartland and Southeast since 1992. Farms in both regions had similar rates of technical advance over the study period.

The use of production contracts continues to be associated with higher farm productivity. The estimates of productivity gains associated with contracting suggest that these productivity advantages helped encourage the recent growth in contracting in the hog industry. Increases in hog farm productivity benefit society through lower food prices for consumers. Productivity gains contributed to about a 30-percent reduction in the price of hogs at the farm gate.

How Was the Study Conducted?

Data used in this report come from USDA surveys of U.S. hog producers conducted for 1992, 1998, and 2004. ERS used a regression analysis to measure hog farm total factor productivity growth between 1992 and 2004 and decompose it into changes in four components: (1) technical change, the increase in the maximum output produced from a given level of inputs; (2) technical efficiency, the farm's ability to achieve maximum output given its set of inputs; (3) scale efficiency, the degree to which a firm optimizes the scale of its operations; and (4) allocative efficiency, a farm's ability to select levels of inputs such that input price ratios equal the ratios of the corresponding marginal products. The study examined variation in economies of scale by farm size, analyzed how increases in scale contributed to productivity growth, and investigated whether scale economies have increased over time. The analysis took advantage of differences in regional growth rates of farm size to examine how limits on the scale of production can affect productivity change. ERS also estimated potential increases in retail pork prices had there been no change in farm productivity.

INTRODUCTION

The increasing size and specialization of hog operations reflect significant structural change in U.S. swine production during the past two decades. Once dominated by small operations that practiced crop and hog farming, the industry has become increasingly concentrated among large operations that produce hogs on several different sites. Further, large operations that specialize in a single phase of production (see glossary) have replaced farrow-to-finish operations that traditionally performed all phases of production. Organizational change in hog production, particularly the widespread use of contracts with growers, has played a key role in the structural shift within the industry by allowing individual producers to grow by specializing in a single phase of production. Technological innovation has also been a driving force behind the structural shift and has contributed to substantial increases in farm productivity.

Changes in the industry have pressured hog producers to adjust the size, organizational structure, and technological base of their operations to remain competitive. The effects of the changes have extended beyond the industry, as restructuring has heightened environmental risks and problems associated with odor, raised concerns about the integrity of rural communities in farming-dependent areas, precipitated controversies over animal welfare, and lowered consumer prices for pork.

This report analyzes changes in the characteristics, production practices, and production costs of U.S. hog operations over the past 15 years. The objective is to emphasize economic relationships that have affected the size and ownership structure of hog production and the impact of these changes on industry productivity.

Data

This report relies on data from detailed surveys of U.S. hog producers for 1992, 1998, and 2004. Data for 1998 and 2004 come from USDA's annual Agricultural Resource Management Survey (ARMS), and data for 1992 come from the ARMS predecessor, USDA's Farm Costs and Returns Survey (FCRS). The surveys cover a cross-section of U.S. hog operations and collect information on size, production costs, business arrangements, production facilities and practices, and farm operator and financial characteristics. The sampling resulted in 1,221 responses from 20 States in 1992, 1,633 responses from 22 States in 1998, and 1,198 responses from 19 States in 2004. Producers in 16 States, including all major hog-producing States, were part of all three surveys in each year (figure 1). These 16 States represented nearly 90 percent of U.S. hog production in each survey year.

Hog farms were chosen from a list of farm operations maintained by USDA's National Agricultural Statistics Service (NASS). The target population of each survey was farms with 25 or more hogs on the operation at any time during the year. The hog producer surveys collect the information necessary to estimate the average cost of production for hog operations. Screening out farms with hog inventories below 25 head excludes farms raising hogs primarily for onfarm consumption and other noncommercial activities, such as youth projects. Each sample included operations with hogs regardless of who owned the hogs, and thus included producers who raised hogs under contract with the hogs' owner. Therefore, results differ from those of surveys of hog owners, such as described by Boessen et al. (2004) (see box, "Hog Producers Versus Hog Owners"). The sample population of hog owners includes very large operations with hogs produced under contractual arrangements on multiple sites.

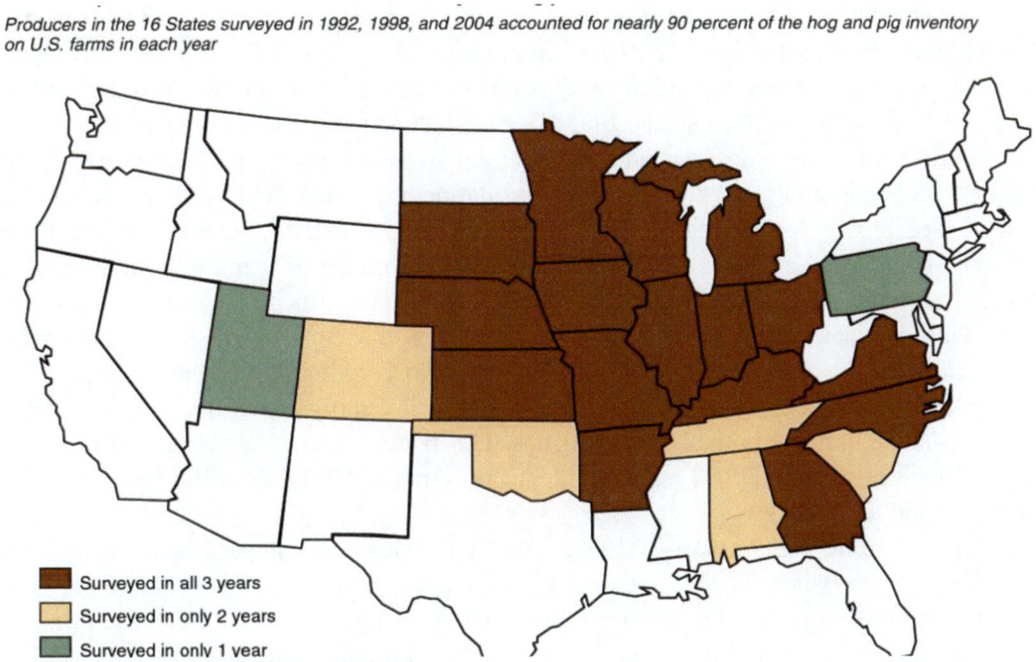

Source: USDA, Economic Research Service.

Figure 1. States as part of the 1992, 1998, and 2004 surveys of hog producers

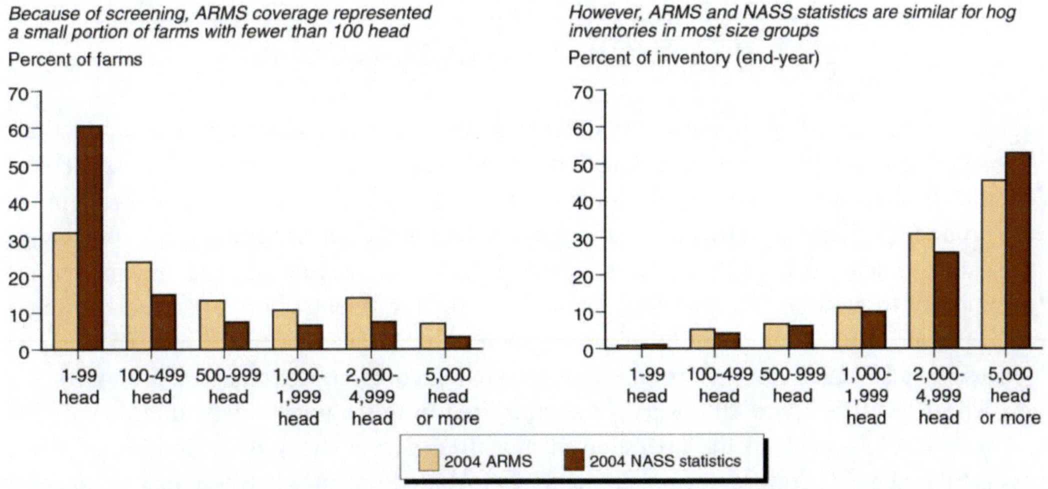

Note: Both the ARMS and NASS estimates are based on surveys of farms with hogs on the operation and thus differences between the estimates are primarily due to the screening out of farms with less than 25 head of hogs in ARMS, and the sampling and nonsampling error in each survey.

Sources: USDA, ERS using data from USDA, NASS, January 2005 and USDA's 2004 Agricultural Resource Management Survey.

Figure 2. Survey coverage of U.S. hog farms and inventory by size of operation, 2004

Each surveyed farm represents a number of similar farms in the population as indicated by the surveyed farm's expansion factor. The expansion factor, or survey weight, was determined from the farm's selection probability and thereby expands the sample to represent the target population. The expanded samples in each survey represent more than 90 percent of the hog and pig inventory on U.S. farms in each survey year (USDA, NASS, July 2005). However, the hog samples expand to cover just over half of the farm operations that had any hogs or pigs (USDA, NASS, January 2005; 1995-99), due to the 25-head threshold.

Figure 2 compares hog farms and inventory by size category from the 2004 ARMS and 2004 NASS hog and pig statistics. Both surveys define hog producers as sites where hogs are located (see box, "Hog Producers Versus Hog Owners"), but the NASS data include all locations with any hogs, and ARMS screens out those with fewer than 25 head. Because of the minimum-size threshold, the share of farms and the share of hog inventory on farms with fewer than 100 head are significantly lower in ARMS than in the NASS statistics. While these small hog operations represent about 60 percent of U.S. hog farms, they include only 1 percent of the hog inventory. The distribution of hog inventory by farm size in the ARMS sample is much like that in the NASS statistics. This pattern of coverage among the different sizes of hog operations in 2004 is nearly identical to that in 1992 (McBride, 1995) and 1998 (McBride and Key, 2003).

Estimates from the surveys in 1992, 1998, and 2004 are comparable because of the consistent way in which the surveys were conducted and processed. Each survey had broad national coverage, represented the same target population (operations with 25 head or more), involved a complex sampling scheme designed to represent the target population, was conducted the same way (hand enumerated) by the same organization (NASS), and collected much the same information in a similar format. Also, the definitions of different types of hog producers used to summarize the data were identical in all years.

Hog Producers Versus Hog Owners

The rapid growth of contract production increasingly has separated hog production from hog ownership. Contract production is an arrangement in which a hog owner (a contractor) engages a producer (a grower) to take custody of the pigs and care for them in the producer's facilities. The producer is paid a fee for the service provided. Contractors typically furnish inputs for growers, provide technical assistance, and assemble the commodity to pass on for final processing or marketing. Contractors often market hogs through marketing contracts or other arrangements with packers or processors. Packers or processors also act as contractors and have production contracts directly with producers.

The 1992, 1998, and 2004 surveys summarized in this report targeted farms with 25 or more hogs located on the operation at any time during the year regardless of who owned the hogs. Therefore, the survey samples included operations where hog producers own their hogs and contract grower operations that are producing hogs owned by a contractor. Contractors are often large conglomerate or corporate organizations that contract with many growers to produce hogs. For example, Smithfield Foods, a packing company, was by far the largest contractor in 2004 with about 800,000 sows (Successful Farming, 2005). ARMS collects information about the hogs owned by contractors such as Smithfield Foods by contacting their contract growers.

The table below shows a comparison of the number of U.S. hog producers versus hog owners and the share of hog inventories of each by size group as reported by USDA's National Agricultural Statistics Service (USDA, NASS, January 2005) for 2004.

The NASS data show about 8,600 fewer hog owners than hog producers (locations with hogs) in 2004, indicating that several hog owners had hogs on multiple operations. The hog inventory was also heavily concentrated among the largest owners, as those with 5,000 head or more owned 75 percent of the inventory. Of these, the largest 110 hog owners, those with 50,000 or more head, owned 54 percent of the inventory. Data reported by Boessen et al. (2004) for 2003 suggest that the very largest hog owners, those with 500,000 or more head, accounted for 40 percent of U.S. hog slaughter, and 23 percent of these hogs were wholly or partially owned by a packer or processor. Study findings by USDA's Grain Inspection, Packers, and Stockyard Administration (USDA, GIPSA, 2007) show that 20 to 30 percent of hogs were owned by packers during 2002-05. The GIPSA study also reports that 89 percent of the total finished-hog volume was marketed through alternatives to the traditional cash or spot market, such as forward contracts, marketing agreements, and packer ownership.

In evaluating the results of these different studies, it is important to recognize that hog industry surveys have different target populations and, hence, provide complementary, rather than duplicate information. GIPSA data are drawn from reports filed by packing plants and are based on the flow of market hogs to slaughter plants. Data reported by Boessen et al. (as well as earlier surveys administered by Lawrence and Grimes) are based on surveys of packers and other hog owners. Owners may be integrators who have their hogs raised on many different contract farms. The ARMS data used here are derived from surveys of sites with hog production facilities, which include farms where hogs are grown under contract for contractors, farms owned by contractors, and independent operations that grow their own hogs and sell them locally or directly to packers.

	Hog producers			**Hog owners**
Head	**Operations**	**Share of inventory**	**Operations**	**Share of inventory**
	Number	*Percent*	*Number*	*Percent*
1-99	42,095	1.0	41,980	1.0
100-499	10,358	4.0	9,800	4.0
500-999	5,155	6.0	3,830	4.5
1,000-1,999	4,449	10.0	2,300	6.0
2,000-4,999	5,137	26.0	1,850	9.5
5,000 or more	2,306	53.0	1,150	75.0
U.S. total	69,500	100.0	60,910	100.0

Source: USDA, ERS using data from USDA, NASS, January 2005.

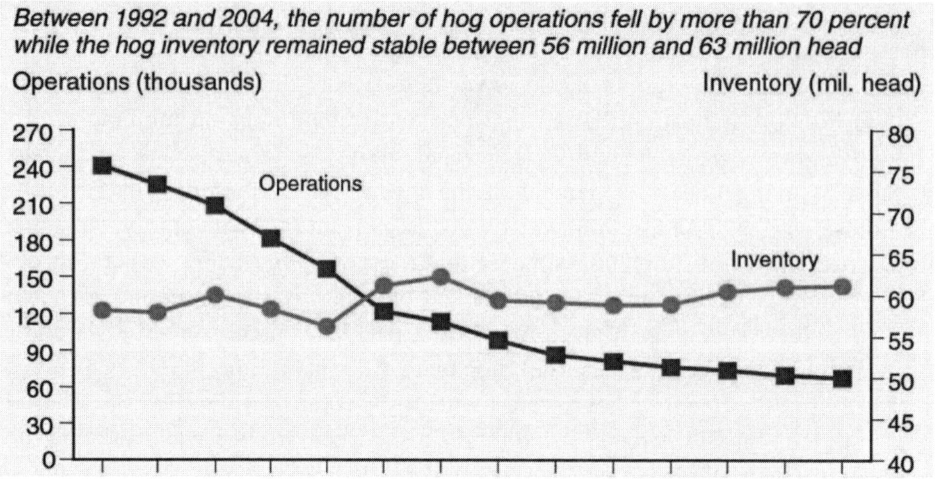

[1] An operation is any place having one or more hogs on hand at any time during the year.
Source: USDA, ERS using data from USDA, NASS, January 2005.

Figure 3. Number of U.S hog operations[1] and hog inventory

Industry Structure and Organization

Changes in the Scale and Approach to Hog Production

The scale of hog production has changed considerably during the past two decades. While the number of farms in the United States remained fairly constant, the number of hog farms fell by more than 70 percent between 1992 and 2004, from over 240,000 to fewer than 70,000 (figure 3). Despite fewer hog farms in the United States, the hog inventory was stable during the period, averaging about 60 million head, with cyclical fluctuations between 56 million and 63 million head (USDA, NASS, 1995-99; January 2005). Thus, hog production consolidated considerably during this period as fewer and larger farms accounted for an increasing share of total output. From 1992 to 2004, the share of the U.S. hog and pig inventory on farms with 2,000 head or more increased from about 30 percent to nearly 80 percent (figure 4). In 2004, farms with 5,000 head or more accounted for more than half of all hogs and pigs.[1]

The sector's approach to hog production also has shifted. An industry characterized by operations that traditionally handled breeding and gestation, farrowing, nursery, and growing-finishing phases (see glossary) of production has given way to one in which operations increasingly specialize in a single phase. Farrow-to-finish units accounted for more than half of hog operations in 1992 but only about a third in 2004 (table 1). Feeder-to-finish operations, specializing in the growing-finishing phase of production, rose from 19 percent of all hog operations in 1992 to 40 percent in 2004.

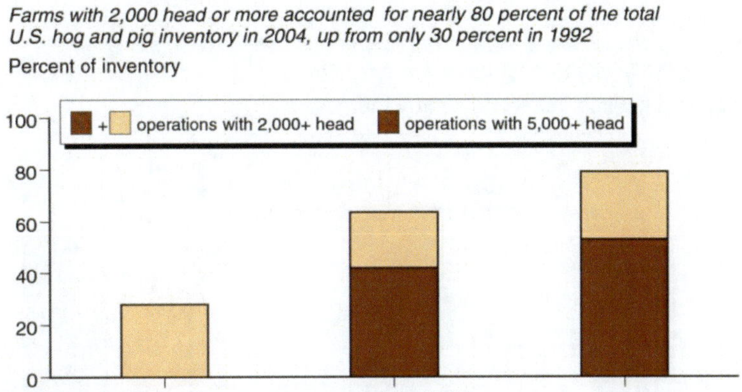

Note: Operations with 5,000+ head were not reported in 1992.
Sources: USDA, ERS using data from USDA, NASS, 1995-99 and January 2005.

Figure 4. U.S. hog and pig inventory on the largest operations

Table 1. Size and structural characteristics by type of hog producer

Item	1992	1998	2004
Farrow-to-finish			
Operations[1] (percent of operations)	54	49	31
Market hogs sold/removed[1] (percent of hogs)	65	38	18
All hog sales/removals (head per farm)	886	1,239	1,472
Contract operations (percent of operations)	id	id	id
Contract production (percent of hogs)	id	id	id
Farm production from hogs (percent of value)	48	47	59
Farm-grown grain fed (percent of feed fed)	55	51	38
Feeder-to-finish			
Operations[1] (percent of operations)	19	31	40
Market hogs sold/removed[1] (percent of hogs)	22	55	77
All hog sales/removals (head per farm)	804	2,756	4,730
Contract operations (percent of operations)	11	34	50
Contract production (percent of hogs)	22	62	73
Farm production from hogs (percent of value)	35	54	72
Farm-grown grain fed (percent of feed fed)	45	22	15
All hog and pig producers			
Operations[1] (percent of operations)	100	100	100
Market hogs sold/removed[1] (percent of hogs)	100	100	100
All hog sales/removals (head per farm)	945	2,589	4,646
Contract operations (percent of operations)	3	15	28
Contract production (percent of hogs)	5	40	67
Farm production from hogs (percent of value)	46	56	71
Farm-grown grain fed (percent of feed fed)	49	35	19

Note: id indicates insufficient data for legal disclosure.
[1] The sum of operations and sales/removals for the producer types will not equal 100 percent because not all producer types are shown.
Source: USDA, ERS using data from USDA's 1992 Farm Costs and Returns Survey and USDA's 1998 and 2004 Agricultural Resource Management Surveys.

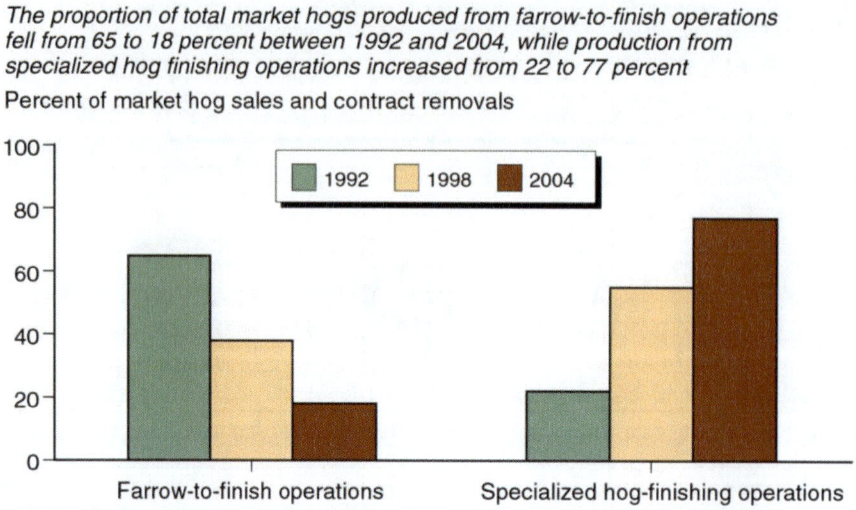

Sources: USDA, ERS using data from USDA's 1992 Farm Costs and Returns Survey and USDA's 1998 and 2004 Agricultural Resource Management Survey.

Figure 5. U.S. hog production by producer type

Changes in the scale of production varied considerably across different types of hog producers. The average size of farrow-to-finish operations grew 66 percent between 1992 and 2004, but specialized hog-finishing operations were more than five times larger in 2004 than in 1992. Consequently, total hog production from specialized operations increased considerably. The share of total U.S. hog production from farrow-to-finish operations fell from 65 to 18 percent between 1992 and 2004, while production from specialized hog-finishing operations increased from 22 to 77 percent of all market hogs sold or removed under contract (figure 5).

Growing numbers of operations specializing in farrowing (farrow-to-wean) and the raising of weanlings (wean-to-feeder) provide further evidence of changing approaches to hog production.[2] These highly specialized operations, rarely reported in surveys prior to 2004, accounted for 7 percent of operations in 2004. They are typically large operations, averaging more than 33,000 and 22,000 head, respectively, sold or removed under contract during 2004. The short time that animals spend in the single production phase of these operations accounts for the large number of hogs that move through during a year. Multiple litters per sow each year also contribute to the high production levels on farrow-to-wean operations.

Changes in the Organization of Hog Production

Changes in production scale and approach have been made possible, in part, by changes in the organizational structure of hog operations as evident in the substantial growth of contract production (see glossary). Production contracts govern the relationship between growers (hog producers) and hog owners ("integrators," or "contractors"), specifying the inputs provided by each party (feeder pigs, feed, labor, capital, energy, transport, and veterinary services and supplies) and the compensation due to each. Contractors typically retain ownership of the hogs on contract operations and compensate growers based on a fee-for-service arrangement, rather than the market price for hogs. Such an arrangement allows individual producers to grow significantly by specializing in one phase of production.

Production contracts differ from marketing contracts, which often govern the relationship between hog owners and hog packers. Marketing contracts specify expected hog quantities and qualities, the location and timing of delivery, and compensation as a hog price or a price formula. The same hog produced under a production contract between a contractor and grower can be sold to a packer under a marketing contract between the contractor and a packer. In this report, contract operations refer to production contracts because the focus is on the growing stage, not on packer procurement (see box, "Hog Producers Versus Hog Owners").

Hog operations with production contracts accounted for only 3 percent of overall U.S. hog operations and 5 percent of U.S. hog production (sales and removals) in 1992 but grew to 28 percent of operations and 67 percent of production (sales and removals) by 2004 (see table 1). Half of feeder-tofinish operations and more than 70 percent of production on feeder-to-finish farms were under production contracts in 2004. Likewise, 67 percent of specialized farrowing operations and more than 90 percent of specialized weanling operations used contractual arrangements. In contrast, contract production was virtually nonexistent on farrow-to-finish farms.

The average size of hog operations increased from 1992 to 2004 but grew the fastest for operations producing under contract (figure 6). Contract feeder pig-to-finish operations averaged about 1,000 more head produced in 1992 than did other operations. By 1998, contract operations averaged 3,700 more head than other operations, and the difference reached 4,500 head in 2004.

Contract hog-finishing operations produced 7,000 head annually on average in 2004, compared with 2,500 head on other operations.

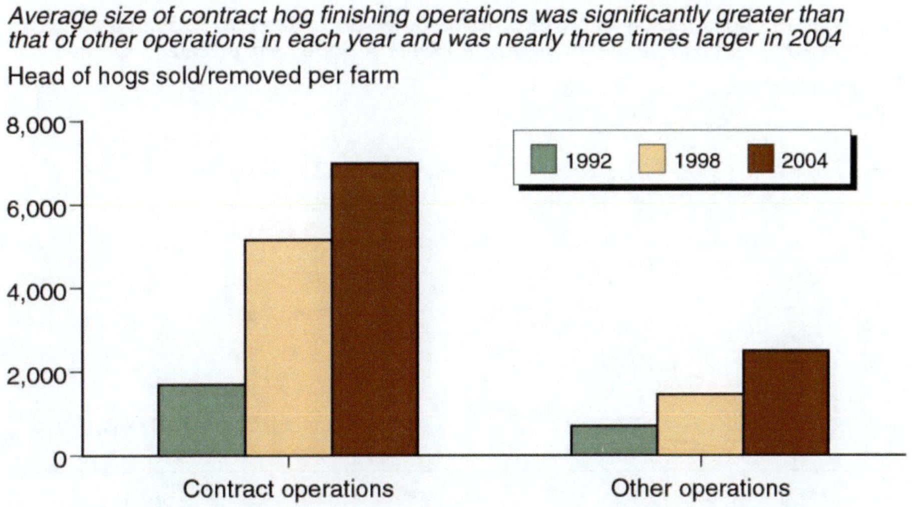

Sources: USDA, ERS using data from USDA's 1992 Farm Costs and Returns Survey and USDA's 1998 and 2004 Agricultural Resource Management Survey.

Figure 6. Size of hog-finishing operations by business arrangement

Among farms with hogs, the average value of farm production from hog enterprises increased from 46 to 71 percent during 1992-2004 (see table 1). That is, hogs generated 71 percent of the total value of farm production on these farms. The share of farm product value from hogs increased most rapidly, from 35 to 72 percent, on hog-finishing operations. Sources of hog feed also indicate increasing specialization in hog production. Grain produced on the same farm accounted for half of the feed consumed by hogs in 1992 but fell below 20 percent by 2004. Again, specialized hog- finishing operations accounted for the fastest change (from 45 to 15 percent). The resulting farms—with greater shares of production value from hogs and more hog feed from off-farm sources—often produce hogs under contract. Under these arrangements, contractors deliver feed from off-farm sources to their growers, allowing individual growers to use their time and financial resources to increase the scale of hog operations rather than expand crop acreage for feed production.

Regional Shifts in Hog Production

Geographical shifts in hog production have accompanied the structural and organizational changes in the industry. Historically, hog production was concentrated in Corn Belt States, where an abundant supply of corn provided a cheap source of hog feed. During the 1980s and 1990s, however, hog production grew dramatically in nontraditional areas, driven mainly by the growth of large contract operations. For example, in North Carolina, the inventory of hogs and pigs more than doubled between 1987 and 1992, as the State's rank in total hog inventory went from sixth to second and then more than doubled again between 1992 and 1998 (figure 7). Since 1992, hog production also has moved aggressively into Western States, where the combined inventory of Oklahoma, Colorado, Texas, and Utah grew from 1.2 million to 4.9 million head between 1992 and 2004.

Hog numbers grew rapidly in North Carolina between 1992 and 1998 but slowly through 2004. Growth has been steady in Western States since 1992.

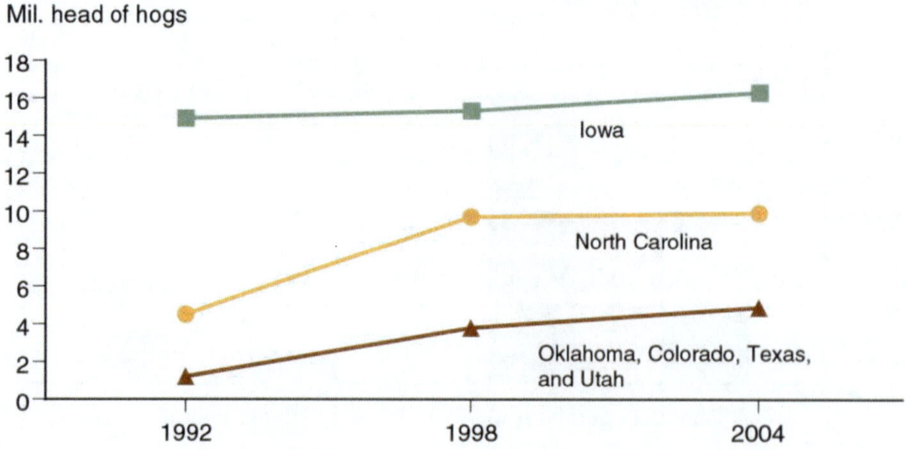

Source: USDA, ERS using data from USDA, NASS, 1995-99.

Figure 7. Hog inventories in selected States

Hog production has traditionally been concentrated in the Heartland, but during the 1980s and 1990s it expanded rapidly in the Southern Seaboard and more recently in western regions, particularly in the Prairie Gateway and Basin and Range

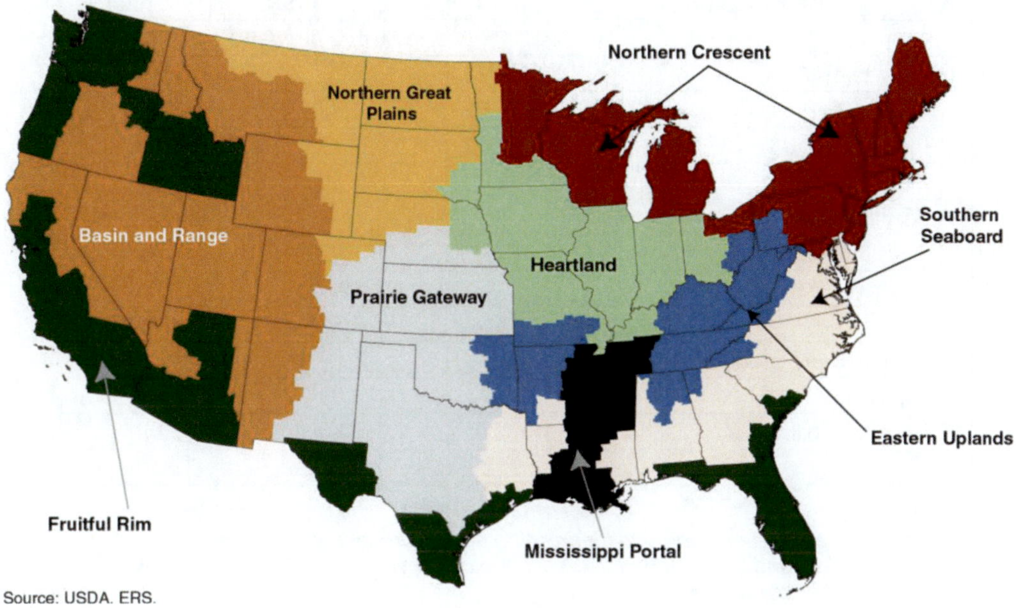

Source: USDA. ERS.

Figure 8. Farm Resource Regions

Rapid growth in the North Carolina hog industry ended after a State law enacted in August 1997 placed a moratorium on the construction of new and expanded hog operations with 250 or more hogs (North Carolina General Assembly, 1997). The purpose of the moratorium, extended for 4 additional years in 2003, was to provide State and local government time to adopt zoning ordinances and gather information on environmental impacts and alternative waste technologies. Restricted growth in North Carolina may explain the particularly rapid growth of the industry in Western States. Open space and a relatively low population density in these States provide greater flexibility in managing animal waste.

ERS has constructed a set of farm resource regions that depict geographic specialization in production of U.S. farm commodities (figure 8). Table 2 shows changes in average production of hog operations during 1992-2004 by farm resource region. The average per farm size of Heartland hog operations grew steadily during the period from just under 1,000 head to more than 5,000 head. In contrast, Southern Seaboard operations increased in average per farm size from 1,200 to over 10,000 head. Most of the growth in Western State operations occurred between 1992 and 1998. Specialized hog-finishing operations accounted for much of the increased operation size in all regions.

Increasing size and specialization of hog-finishing operations accompanied rapid growth in regional contract production during this period. The share of hogs finished under contract in the Heartland increased from only 4 percent in 1992 to 26 percent in 1998 and to almost 50 percent in 2004 (figure 9). Contract production in the Southern Seaboard increased from 12 to 85 percent between 1992 and 1998 and accounted for virtually all market hogs produced in 2004. The 1992 survey did not measure any contract production of market hogs in the Western States but by 2004, nearly half of market hogs were produced under contract.

Table 2. Regional hog and pig sales and contract removals by type of hog producer

Item/Region	1992	1998	2004
	Average head per farm		
Farrow-to-finish			
Heartland	901	1,288	1,851
Southern Seaboard	1,093	1,163	1,068
West	621	1,305	1,459
Feeder-to-finish			
Heartland	833	1,972	4,152
Southern Seaboard	1,035	10,951	12,057
West	358	3,589	3,255
All hog and pig producers			
Heartland	975	2,098	5,106
Southern Seaboard	1,206	10,021	13,995
West	702	2,231	1,859

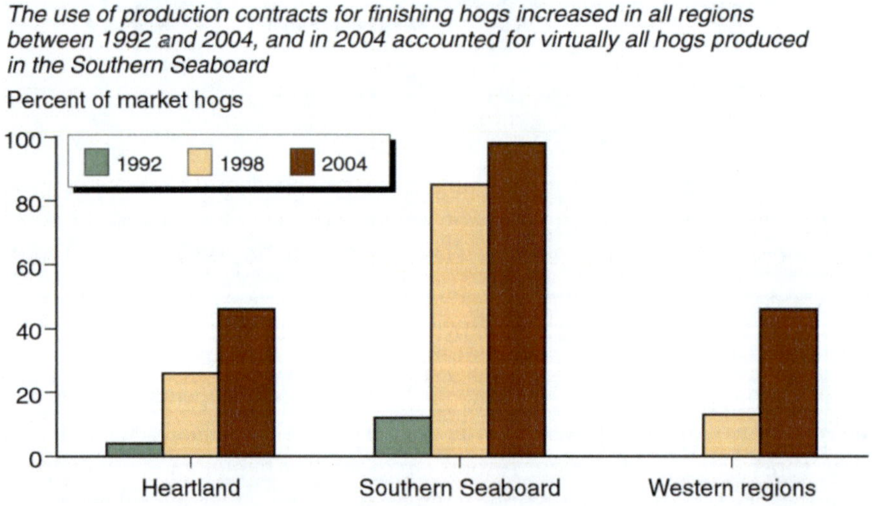

Sources: USDA, ERS using data from USDA's 1992 Farm Costs and Returns Survey and USDA's 1998 and 2004 Agricultural Resource Management Survey.

Figure 9. Market hogs removed under contract by region

Technological Innovation in Hog Production

Like contract production, technological innovation has facilitated change in hog production. Technological innovation in hog production includes such advances as improved genetics, nutrition, housing and handling equipment, veterinary and medical services, and management that improves the performance of hogs and the efficiency of the operation and/or reduces production risk.

Data from USDA's National Animal Health Monitoring Service (NAHMS) surveys conducted in 1990, 1995, and 2000 show technological innovation on hog operations with 100 or more head (USDA, APHIS, 2005). For example, artificial insemination (AI) improves the genetic potential of the swine herd and the conception rates of breeding animals. Between 1990 and 2000, the share of hog operations using AI increased from 7 to 23 percent, while the share of sows and gilts mated by AI increased from 1 to 73 percent. Another innovative practice to enhance productivity, all-in/all-out housing management, commingles pigs of a similar age and weight and keeps the entire group together as it moves through each production phase. The hogs are marketed a room at a time, and rooms are washed and disinfected after each group leaves. The NAHMS data revealed that the use of all-in/all-out management for finishing hogs increased from 25 percent of hog operations in 1990 to 57 percent in 2000; by 2000, about 85 percent of finished hogs were managed with all-in/all-out systems.

The 2004 ARMS collected information about use of AI, all-in/all-out management, and other technologies, including terminal crossbreeding programs, commercial seed stock obtained from high-quality breeding animals, and phase feeding, which varies feed to match animal diets with changing nutritional requirements. Analysis of 2004 ARMS data reveals that specialized hog producers were more likely than farrow-to-finish operations to use these practices. Farrow-to-wean operations more often used AI, terminal crossbreeding, and commercial seed stock than did farrow-to-finish operations (figure 10).

Feeder-to-finish operations more often used phase feeding and all-in/all-out management of finishing facilities (figure 11). There also appears to be a strong relationship between the use of these practices and the size of an operation (table 3). AI was used on only 4 percent of the smallest farrow-to-finish operations in 2004 but on 92 percent of the largest. As farrow-to-finish operations and feeder-to-finish operations increased in size, use of all-in/all-out finishing increased from 14 to 83 percent of farms and from 66 to 92 percent of farms, respectively. Because large, specialized hog operations can spread fixed costs over more production and more easily take advantage of resulting productivity gains, they are better able to invest in innovative hog-production technologies.

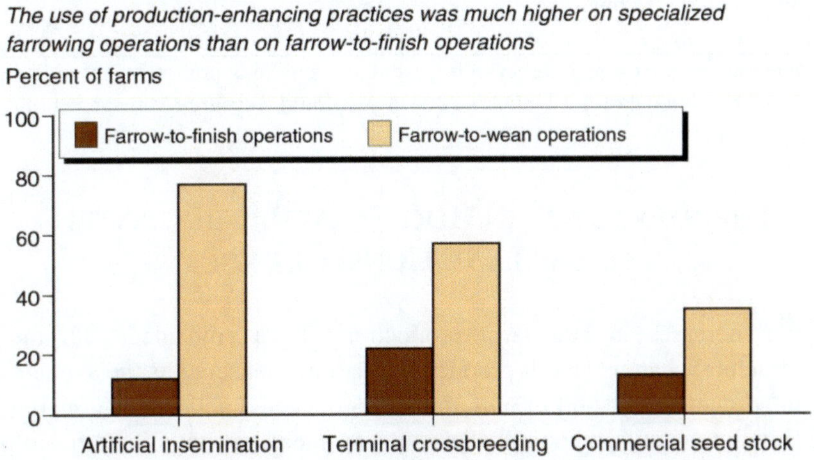

Source: USDA, ERS using data from USDA's 2004 Agricultural Resource Management Survey.

Figure 10. Practices used by operations farrowing pigs, 2004

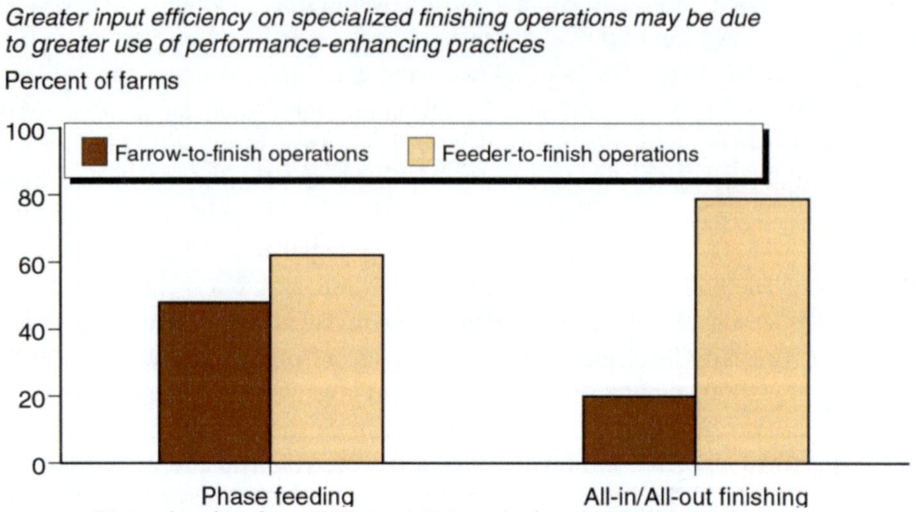

Source: USDA, ERS using data from USDA's 2004 Agricultural Resource Management Survey.

Figure 11. Practices used by operations finishing hogs, 2004

Table 3. Production practice use by size and type of hog producer, 2004

	Size of operation[1]			
Item	**Fewer than 500 head**	**500-1,999 head**	**2,000-4,999 head**	**5,000 head or more**
	Percent of farms			
Farrow-to-finish operations				
Artificial insemination	4	12	51	92
Terminal crossbreeding	11	38	43	73
Commercial seed stock	5	24	36	26
Phase feeding	42	53	61	84
All-in/all-out finishing	14	20	54	83
Feeder-to-finish Phase feeding	51	60	72	72
All-in/all-out finishing	66	80	86	92

[1] Size of operation is the maximum number of hogs and pigs on the operation at any time during 2004.
Source: USDA, ERS using data from USDA's 2004 Agricultural Resource Management Survey.

Improvements in Hog Farm Productivity: Causes and Consequences

Substantial increases in hog farm productivity have coincided with the industry's pronounced structural changes. Farm productivity can be measured as the average quantity of inputs used in production per unit of output. ARMS collected detailed information about hog production, including inputs and output. This analysis measures output in terms of *hog weight gain*—the weight added during the prior calendar year to purchased/placed hogs that were later sold/removed, plus the total weight added to the hog inventory. Hog weight gain, unlike the alternative output measure "number of head removed," accounts for changes in inventory and for differences in the weights of feeder and finished pigs across operations.[3] ARMS asked farmers about changes in their hog inventory and about the quantity and weights of hogs moved on and off the farm. The survey also asked farmers about the amount and types of feed purchased by the operation or contractor, the amount of homegrown feed used, and the number of hours of labor spent on the hog enterprise (including time by the operator and paid and unpaid labor).

The average quantity of feed required per hundredweight of gain declined 14.9 percent (1.3 percent average annual decline) for farrow-to-finish operations between 1992 and 2004 and declined 44.1 percent (4.7 percent annually) for feeder-to-finish operations (table 4). The average quantity of labor used per hundredweight declined even more—falling 52.5 percent (6.0 percent annually) for farrow-to-finish operations from 1992 to 2004, and falling 83.1 percent (13.8 percent annually) for feeder-to-finish operations.

Productivity gains contributed to a decline in production costs between 1992 and 2004. For all farrow-to-finish hog producers, average production costs per hundredweight of gain, expressed in 2004 dollars, were 27.9 percent lower in 2004 than in 1992. This change amounts to a 2.7-percent average annual rate of decline. Real costs declined faster for feeder-to-finish hog producers, falling 43.9 percent between 1992 and 2004, or 4.7 percent annually.

Sources of Productivity Growth

Economic competition and the incentive to maximize profits drive structural changes in the hog industry. If larger operations are more profitable than smaller ones, competitive pressures may be expected to result in a larger average farm size in the long run. Similarly, operations that are first to adopt a cost-saving technology, in regions with lower input costs, or closer to markets, have a competitive advantage that makes them more likely to survive and grow. Relationships between farmers and processors also evolve to reflect more cost-effective modes of production. Since 1992, the use of production contracts has increased dramatically at the expense of independent production. The organizational structure of the industry also reflects efficiency gains from increased specialization of the various phases of hog production on separate operations.

Given output and input prices, the total factor productivity (TFP) of the farm determines returns. TFP is the quantity of farm output per unit of inputs (with all inputs aggregated). TFP reflects the production technology available (which determines the rate at which inputs can be combined to make outputs), whether farms are operating at an efficient scale of production, how efficiently inputs are combined given the production technology, and how well the farmer takes into account the relative prices of inputs.

The production technology used by a farm is a fundamental determinant of its productivity. In hog production, the production technology incorporates livestock genetics, feed mixtures and feed equipment, housing and handling equipment, and veterinary and medical services used. The term *technical change* (or *progress*) describes the increase in productivity resulting from adopting more efficient production technologies.

An increase in the scale of production is another source of productivity growth when there are increasing returns to scale—that is, when a proportional increase in all inputs results in a more than proportional increase in output. The returns to scale of a particular production technology are measured by its "scale elasticity"—the percentage increase in output obtained from a 1-percent increase in all inputs.[4] The movement toward the optimal scale of production (the scale at which the scale elasticity equals one) is said to increase *scale efficiency.*

Table 4. Efficiency and production costs per cwt by type of hog producer

Item	1992	1998	2004
Farrow-to-finish			
Feed conversion rate (lbs per cwt gain)	416	374	354
Labor rate (hrs per cwt gain)	1.13	0.72	0.54
Production costs, current dollars[1]	46.63	43.50	42.44
Production costs, 2004 dollars[1]	58.89	49.20	42.44
Feeder-to-finish			
Feed conversion rate (lbs per cwt gain)	383	282	214
Labor rate (hrs per cwt gain)	0.89	0.24	0.15
Production costs, current dollars[1]	37.54	31.08	26.59
Production costs, 2004 dollars[1]	47.41	35.15	26.59

[1]Production costs are the sum of operating and capital costs less costs for feeder and nursery pigs. Pig costs are excluded because they are not an input contributing to weight gain. 1992 and 1998 costs are deflated to 2004 dollars using the national GDP implicit price deflator (Bureau of Economic Analysis).

Source: USDA, ERS using data from USDA's 1992 Farm Costs and Returns Survey and USDA's 1998 and 2004 Agricultural Resource Management Surveys.

Table 5. Production costs per cwt by type and size of hog producer

Item	1992	1998	2004
		Dollars per cwt gain	
Farrow-to-finish			
Cwt gain<1 ,000	72.38	78.39	73.55
1 ,000≤cwt gain< 2,500	63.26	57.70	51.29
2,500≤cwt gain<1 0,000	54.88	46.91	39.94
10,000≤cwt gain<25,000	54.12	39.35	37.52
25,000≤cwt gain	id	38.61	36.03
Feeder-to-finish			
Cwt gain<1 ,000	61.99	57.49	45.46
1 ,000≤cwt gain< 2,500	46.07	48.02	33.34
2,500≤cwt gain<1 0,000	43.70	36.03	31.03
1 0,000≤cwt gain<25,000	id	26.97	23.03
25,000≤cwt gain	id	26.26	24.06

Note: Production costs are the sum of operating and capital costs less costs for feeder and nursery pigs. Pig costs are excluded because they are not an input contributing to weight gain. 1992 and 1998 costs are deflated to 2004 dollars using the national GDP implicit price deflator (Bureau of Economic Analysis).

id = insufficient data for legal disclosure.

Source: USDA, ERS using data from USDA's 1992 Farm Costs and Returns Survey and USDA's 1998 and 2004 Agricultural Resource Management Surveys.

Table 5 suggests increasing returns to scale in each survey year—per unit production costs decline as the scale of production increases.[5] It seems likely that some of the decrease in average unit costs shown in table 4 has resulted from the growth in the size of operations. That is, between 1992 and 2004, some farms responded to the economic incentive to reduce average costs by expanding the scale of their operations, some smaller, less- efficient operations exited the industry, and new operations entered at a larger, more efficient scale.

While increases in scale efficiency and technological change are likely the largest sources of productivity growth, farms also may become more productive by increasing technical and allocative efficiency. Holding the scale of production and the technology constant, *technical efficiency* increases if farmers use inputs more efficiently in the production process. For example, a farm manager might increase technical efficiency by carefully blending the contents of feed to maximize animal weight gain per unit of feed. The farmer does not use a new technology or produce more, but the productivity of the farm increases because input expenditures per unit decline.

Farmers increase *allocative efficiency* if they can improve productivity by choosing the mix of inputs to better reflect their relative costs. For example, if the price of feed increases relative to the price of capital, then it becomes more efficient to substitute capital for feed (say, by using machinery that more accurately rations feed).

This study focuses on measuring and understanding changes in *farm-level* total factor productivity. It uses ARMS data to disaggregate changes in farm TFP into changes in four constituent parts: technical change, technical efficiency, scale efficiency, and allocative efficiency. Productivity gains at the industry level have accompanied substantial farm-level productivity gains. Industry-level efficiency gains result in part from increasing specialization in the various stages of hog production. As discussed in the previous chapter, the hog industry has seen finished hog production move from one to two or three separate operations. That is, since 1992, the share of hog output produced by farrow-to-finish operations has declined, while the share of output produced by feeder-to-finish operations has increased. Feeder-to-finish operations obtain feeder pigs from separate operations that raise feeder pigs (farrow-to-feeder or farrow-to-wean and wean-to-feeder). As production shifted from less efficient farrow-to-finish operations to more efficient specialized operations, the total costs of producing finished hogs industrywide may have declined substantially, resulting in industry-wide improvements in productivity. This study does not assess industrywide efficiency gains from changes in specialization.[6] Instead, it examines productivity growth at one important stage of hog production—feeder-to-finish —which now accounts for most finished hog output (having increased from 22 percent to 77 percent of all output between 1992 and 2004 (see table 1).[7] Feeder-to-finish operations make an interesting case study because their large and increasing share of output suggests these farms will dominate the industry in the future, and they have had the fastest productivity growth since 1992 (see table 4) and the greatest increase in scale of production (see table 1).

Disaggregating Productivity Growth

Disaggregating the observed increases in total factor productivity into technical change, technical efficiency change, scale efficiency change, and allocative efficiency change provides insights into the forces that drive structural change. The methodology used here to disaggregate TFP follows Orea (2002) and is described in more detail in the appendix. The approach requires estimation of a production frontier—a parametric relationship between input quantities and the maximum output achievable from those inputs. The frontier describes the amount that technically efficient operators could produce if they used the best practices available in the industry. Since no producers are perfectly technically efficient, production occurs within the frontier.

The assumed functional relationship between the inputs and output is "flexible" in that it imposes few *a priori* restrictions on the characteristics of the production technology, such as constant returns to scale. The parameters describing the frontier are estimated using a maximum likelihood technique that accounts for the facts that: (1) we do not observe the distance of the actual production levels from the frontier, and (2) input and output levels are measured with error (Battese and Coelli, 1992).

Economists calculate an index of technical efficiency as a farm's ratio of observed output to what could be produced with the same inputs if the farm operated on the estimated production frontier. If the same farm could be observed over time, changes could be tracked in this index (a measure of technical efficiency). However, because ARMS is a repeated cross-section rather than a panel, the study estimates technical efficiency change for a representative (average) farm.

The estimation method permits the parameters of the production frontier to vary over time to allow for technical change. Technical change measures output changes resulting from changes in production technology, holding efficiency, scale, and prices constant. Scale efficiency change captures TFP changes caused solely by changes in input levels, holding technical efficiency, the production function, and prices constant.

Allocative efficiency change measures TFP changes resulting from changes in the "cost effectiveness" of inputs. Relative to their contributions to output, some inputs may be relatively "expensive," and others may be relatively "cheap." Allocative efficiency improves when a firm uses more of the relatively inexpensive inputs, and less of the relatively expensive inputs. TFP is an index of output produced per unit of inputs, where inputs are aggregated into an index using prices as weights. Allocative efficiency change is the residual difference between the total change in TFP (which depends on prices) and the change in the TFP explained by changes in production technology, efficiency, and scale (changes that do not depend on prices).

By definition, the percentage change in TFP equals the sum of technical change plus changes in technical efficiency, scale efficiency, and allocative efficiency. For all farms, TFP more than doubled between 1992 and 2004, with an average increase of 6.3 percent per year (table 6). This is a very high rate of growth—about three times the historical growth rate in productivity for the agricultural sector as a whole (Ahearn et al., 1998). Average annual TFP growth rates were similar in both periods between the ARMS surveys—with average increases of 6.4 percent from 1992 to 1998 and 6.3 percent from 1998 to 2004.

Technological change and increases in scale efficiency accounted for most of the growth in TFP (increasing 3.0 percent and 3.4 percent annually, respectively). The contribution of technological change to productivity growth increased substantially in the second period—from 2.1 percent annually between 1992 and 1998 to 3.9 percent annually between 1998 and 2004. In contrast, the scale effect diminished over time: scale efficiency increased 4.5 percent annually between 1992 and 1998 but only 2.2 percent between 1998 and 2004.

Average technical efficiency changed little over the 12-year study period. Allocative efficiency change also played a small role in TFP change— increasing at an annual rate of 0.5 percent.

Table 6. Decomposition of total factor productivity change (feeder to finish)

Item	Factor change (Annual growth rate)		
	1992-98	1998-2004	1992-2004
	Percent		
Technical efficiency	-1.7	0.8	-0.9
	(-0.3)	(0.1)	(-0.1)
Technology	13.5	25.6	42.5
	(2.1)	(3.9)	(3.0)
Scale efficiency	30.6	13.8	48.6
	(4.5)	(2.2)	(3.4)
Allocative efficiency	2.6	3.9	6.7
	(0.4)	(0.6)	(0.5)
Total factor productivity	45.1	44.1	109.1
	(6.4)	(6.3)	(6.3)

Source: USDA, ERS using data from USDA's 1992 Farm Costs and Returns Survey and USDA's 1998 and 2004 Agricultural Resource Management Surveys.

Table 7. Share of output and scale elasticity by farm size and year (feeder to finish)

Farm size category	1992	1998	2004
	Share of total output (percent)		
Cwt gain < 1,000	14.7	1.9	0.5
1,000 ≤ cwt gain < 2,500	35.0	6.7	3.0
2,500 ≤cwt gain < 10,000	41.0	26.5	16.7
10,000 ≤ cwt gain < 25,000	9.3	29.2	36.3
25,000 ≤ cwt gain	id	35.7	43.4
	Scale elasticity		
Cwt gain < 1,000	1.20	1.24	1.27
1,000 ≤ cwt gain < 2,500	1.13	1.16	1.22
2,500 ≤cwt gain < 10,000	1.08	1.12	1.17
10,000 ≤ cwt gain < 25,000	1.07	1.09	1.12
25,000 ≤ cwt gain	id	1.03	1.05
All farms (mean)	1.16	1.12	1.14
All farms (output-weighted mean)	1.10	1.05	1.07

id = insufficient data for legal disclosure.

Source: USDA, ERS using data from USDA's 1992 Farm Costs and Returns Survey and USDA's 1998 and 2004 Agricultural Resource Management Surveys.

Implications for Scale of Production

Increases in scale efficiency contributed significantly to productivity gains between 1992 and 2004 as farms grew in size to take advantage of increasing returns to scale. Estimates of returns to scale provide insight into farmers' incentives to undertake further increases in scale.

The top of table 7 shows the change over time in the share of total finished- hog output produced by farms in each farm-size category. The increase in share was particularly notable for large operations. For example, so few farms were producing more than 25,000 hundredweight gain that they were not even sampled in the 1992 survey. By 1998, these operations produced 35.7 percent of total output, and the share rose to 43.4 percent by 2002.

The bottom of table 7 reports two estimates of average scale elasticity for all farms in each survey year—the mean and weighted mean. The mean scale elasticity for all farms ranges between 1.12 and 1.16, which indicates that a 10-percent increase in inputs produces an 11.2- to 11.6-percent increase in output for the "typical" farm. The mean weighted by farm output is the scale elasticity associated with the "typical" quantity of output. The weighted mean is smaller than the unweighted mean because larger farms produce more output and the scale elasticity declines with size of the operation. Both the mean and weighted mean values imply increasing returns to scale in all periods.

As expected, scale elasticity declines as farm size increases—large farms obtain smaller gains from increasing scale than do small farms. The technology used by farms in the largest category exhibits positive returns to scale. However, the potential for efficiency gains from further increases in scale appear limited for large farms—farms producing more than 25,000 hundredweight had an average scale elasticity of 1.05 in 2004. On the other hand, potential scale efficiency gains remain in the sector as a whole—farms producing less than 25,000 cwt accounted for more than half of all output in 2004.

Returns to scale increased in all size categories between 1992 and 1998 and between 1998 and 2004. This implies that holding output constant (output is approximately constant within each category), returns to scale increased steadily over time. Hence, the reduction in the contribution of scale efficiency to TFP (see table 6) results from a slowdown in the growth of average farm output (see table 1), not from a reduction in the optimal scale of production.[8] Because average farm size increased substantially over the study period, the average scale elasticity for all farms showed little change.

Regional Differences in Productivity Growth

Productivity growth in the U.S. hog sector varied substantially by region between 1992 and 2004 (table 8). For feeder-to-finish farms, this study focuses on two major hog-producing regions: the Heartland (IA, IL, IN, KY, MO, and OH) and the Southeast (AL, AR, GA, NC, SC, and VA). Producers in the remaining surveyed States (CO, KS, MI, MN, NE, OK, PA, SD, TN, TX, UT, and WI) are placed in the "other regions" category.

Table 8. Summary statistics by region (feeder to finish)

Item	1992	1998	2004
		Percent	
Share of feeder-to-finish farms			
Heartland	54.7	55.9	48.9
Southeast	15.2	9.6	10.7
Other regions	30.1	34.5	40.4
Share of feeder-to-finish output			
Heartland	57.9	35.4	45.2
Southeast	20.1	32.3	24.7
Other regions	22.0	32.3	30.0
		Hundredweight gain	
Mean farm output			
Heartland	1,716	5,399	11,313
Southeast	2,333	20,771	25,074
Other regions	1,097	10,516	12,933

Source: USDA, ERS using data from USDA's 1992 Farm Costs and Returns Survey and USDA's 1998 and 2004 Agricultural Resource Management Surveys.

Production shifted from the Heartland to the Southeast and other regions during 1992-98. The share of output produced by farms in the Southeast increased 12.2 percentage points, even though the share of feeder-to-finish operations located in this region declined 5.6 percentage points. A large increase in scale of production accounts for this increase in output share despite a decline in share of farms: average farm size in the Southeast increased almost tenfold. Farms in the Heartland, while representing roughly half of all feeder-to-finish hog farms in both 1992 and 1998, experienced smaller proportional increases in average farm output over this period, and, consequently, their share of total output declined by 22.5 percentage points.

During 1998-2004, feeder-to-finish output share rebounded in the Heartland and declined in the Southeast. Heartland farms doubled in size while farms in the Southeast had much smaller proportional increases (though starting from a larger average size). As a result, the Heartland increased its share of output 9.8 percentage points over this period, while the Southeast decreased its share by 7.6 percentage points.

The decline in the output share and in the rate of growth in average farm size in the Southeast during 1998-2004 can probably be attributed largely to policy changes at the State level. Over the three survey periods, farms in North Carolina produced on average about 92 percent of total output in the Southeast region. In 1997, North Carolina passed the Clean Water Responsibility and Environmentally Sound Policy Act, which imposed a moratorium on the construction of new and expanded hog operations with 250 or more head. The moratorium contained several exceptions, including new construction using "innovative animal waste management systems that do not employ an anaerobic lagoon."[9] Though the moratorium was originally set to expire in 1999, North Carolina extended it several times in modified form through 2007.

Except for "other inputs" in the Southeast, all partial factor productivity measures in the three regions increased at similar annual rates between 1992 and 2004 (table 9). However, this pattern masks substantial differences in factor productivity between the Heartland and the Southeast during the two subperiods. While each of the three regions began in 1992 with approximately the same levels of factor productivity, from 1992 to 1998, farms in the Southeast experienced much larger increases in feed, labor, and capital productivity than did farms in the Heartland. Between 1998 and 2004, this pattern reversed, with farms in the Heartland increasing their feed, labor, and capital productivity more rapidly than farms in the Southeast.

Disaggregating the change in productivity for each region shows the extent to which changes in the scale of production, or differences in rates of technological change, allocative efficiency change, or technical efficiency change caused these shifts in productivity (table 10).[10] The regional changes in TFP are consistent with changes in partial factor productivity previously discussed (see table 9).

Table 9. Partial factor productivity by region and year (feeder to finish)

Input/Region	Partial factor productivity			
	1992	1998	2004	Annual growth rate, 1992-2004
Feed productivity (cwt gain per cwt feed)				
Heartland	0.286	0.314	0.764	8.5
Southeast	0.281	0.443	0.629	6.9
Other regions	0.243	0.313	0.625	8.2
Labor productivity (cwt gain per unit of hog enterprise labor[1])				
Heartland	2,070	3,019	6,187	9.6
Southeast	2,237	6,151	6,918	9.9
Other regions	2,584	2,919	5,373	6.3
Capital productivity (cwt gain per dollar[2])				
Heartland	0.091	0.097	0.238	8.3
Southeast	0.099	0.156	0.252	8.1
Other regions	0.075	0.111	0.234	9.9
Other inputs productivity (cwt gain per dollar[3])				
Heartland	0.327	0.491	0.541	4.3
Southeast	0.456	0.359	0.485	0.5
Other regions	0.248	0.491	0.490	5.8

[1]The labor input is a weighted index (Tornqvist index) of paid labor plus unpaid farm household labor that uses the labor expenditure shares for paid and unpaid labor as weights. The labor expenditures for paid labor are observed. Labor expenditures for unpaid labor are estimated using an imputed wage for unpaid labor.

[2]Capital is the "capital recovery cost"—the estimated cost of replacing the existing capital equipment (barns, feeding equipment, etc.).

[3]Other inputs are defined as real expenditures on veterinary services, bedding, marketing, custom work, energy, and repairs.

Source: USDA, ERS using data from USDA's 1992 Farm Costs and Returns Survey and USDA's 1998 and 2004 Agricultural Resource Management Surveys.

Table 10. Decomposition of total factor productivity change (feeder to finish)

	1992-98	1998-2004	1992-2004
Heartland			
Technical efficiency	-0.5	0.2	-0.2
Technology	2.2	3.9	3.0
Scale efficiency	3.1	4.4	3.7
Allocative efficiency	0.9	0.6	0.8
Total factor productivity	5.3	8.1	6.7
Southeast			
Technical efficiency	0.1	-0.6	-0.3
Technology	2.3	4.4	3.4
Scale efficiency	9.0	2.2	5.5
Allocative efficiency	1.4	-0.7	0.4
Total factor productivity	11.5	5.2	8.3
Other regions			
Technical efficiency	0.1	0.2	0.1
Technology	2.1	3.7	2.9
Scale efficiency	5.6	-1.5	2.0
Allocative efficiency	-0.7	1.1	0.2
Total factor productivity	6.7	3.6	5.2

Source: USDA, ERS using data from USDA's 1992 Farm Costs and Returns Survey and USDA's 1998 and 2004 Agricultural Resource Management Surveys.

The average annual growth rates imply that between 1992 and 1998, TFP almost doubled in the Southeast but increased by only about a third in the Heartland over the same 6-year period. Between 1992 and 1998, technical progress contributed roughly equal amounts to the growth in TFP for farms in both the Heartland and the Southeast regions. However, the contribution of scale efficiency to TFP was much greater in the Southeast than in the Heartland (increasing annually at 9.0 percent versus 3.1 percent). The large increase in scale efficiency in the Southeast resulted from the region's rapid increase in the scale of production (see table 8), given the increasing returns to scale of the production technology.

Between 1998 and 2004, productivity in the Heartland rebounded— increasing by almost 60 percent, compared with only 36 percent in the Southeast. Increases in scale efficiency drove faster growth in productivity in the Heartland in the second period—scale efficiency increased at an annual average rate of 4.4 percent, compared with only 2.2 percent in the Southeast. The Heartland actually lagged slightly behind the Southeast in technological progress during this period.

In sum, average farm-size growth and the resulting improvements in scale efficiency explain most differences in productivity growth between the Heartland and Southeast since 1992. While the rate of technical change approximately doubled between the periods in each region, farms in all regions had similar rates of technical advance. This suggests that the adoption of new technologies, information, and genetic improvements diffused at similar rates across the Nation. Because farms in the Heartland operate at a smaller average scale than do farms in the Southeast, the Heartland retains greater scope for further scale efficiency gains.

Organizational Structure and Productivity

As documented earlier, the use of production contracts increased dramatically among feeder-to-finish hog operations since at least 1992 (see table 1). Production contracts offer several potential advantages over independent production that help explain their growing use: contracts can reduce information asymmetries between growers and processors, improve coordination and timing of product delivery, and lower income risk for growers. Production contracts also may raise farm productivity by improving the quality of farm management decisions, speeding the transfer of technical information to growers, improving growers' access to credit, and facilitating the adoption of more efficient technologies.

Recent ERS research showed a link between the use of production contracts and hog farm productivity. Using the 1998 ARMS survey of feeder-to-finish hog farms, Key and McBride (2003) compared the productivity of similar independent operations and contract operations, controlling for unobservable differences that might be associated with the decision to contract. The authors found that production contracts were associated with an average increase in total factor productivity of about 23 percent.

Given the scope of structural changes in the industry since 1998, it is possible that the observed differences in productivity between contract and independent operations have diminished. Many less-efficient independent operations have exited the industry, and some independent operations have begun to contract. Key and McBride (2007) used data from the 2004 ARMS to examine whether production contracts remain associated with greater farm productivity. The authors used an instrumental variables technique to isolate the effect of contracts on productivity. As in the earlier study, the authors found that contract operations were substantially more productive than similar independent operations. A 10-percent increase in the prevalence of contracting would increase average total factor productivity by 5 percent.

The estimates of the magnitude of the productivity gains attributable to contracting suggest that these productivity advantages contributed to the recent growth in contracting in the hog industry. The apparent continuing link between contracts and productivity, along with lower grower income risk and improved coordination and timing of product delivery for contractors, suggest that the use of production contracts is likely to continue to expand.

The connection between contracts and productivity suggests that contracting may have played a role in the recent increase in the average scale of production. Because contract operations are larger operations on average, it will be larger operations that enjoy the productivity gains from contracting. Consequently, contracting may enhance the competitive position of larger producers vis-à-vis smaller producers.

Implications of Productivity Gains for Consumers

A main potential benefit to society of increases in hog farm productivity is lower food prices for consumers. How much have productivity gains in hog production been reflected in retail prices? One way to address this question is to estimate how much hog prices would have increased had there been no change in farm productivity. This counterfactual can be estimated by examining input prices. In a competitive market, the price received by farmers for finished hogs equals the total cost of inputs plus a "normal" rate of return. Consequently, if the normal rate of return were constant and farm productivity did not change, then hog prices would be expected to track input prices.

Figure 12 shows price trends for the major inputs to hog production. Each input price index is scaled so that the index at the beginning of the series (January 1992) equals the longrun average input cost share estimated using ARMS data. Specifically, at the beginning of the series, input costs comprised feed costs (55 percent), capital (20 percent), labor (17 percent), and other inputs (8 percent). The total input cost index is the sum of the input price indices, which equals 1.0 by definition in January 1992. The figure shows that feed costs, which fluctuated between 51 and 73 percent of total costs, drive the variation in the total cost index. A linear time trend fitted to the total cost index (heavy black line) indicates that input prices increased 28.5 percent over the study period. Since hog prices reflect the costs of production on the farm, this result implies that the gross farm value would have increased by about 28.5 percent between 1992 and 2004 if total factor productivity had not changed.

Figure 13 illustrates price trends at the farm, packer, and retail levels. The gross farm value is the value of the hog when it is sold, measured in cents per pound of retail weight. The wholesale value is the average value of the meat as it leaves the packing plant, measured in cents per pound of retail weight. The retail value is the average value of selected cuts of meat at the grocery store, measured in cents per pound of retail weight. The heavy black line shows an estimated linear time trend in the gross farm value of hogs. The estimated trend indicates that hog prices declined 3.3 percent between 1992 and 2004.

To summarize, input price increases imply that finish hog prices would have risen by 28.5 percent with no change in productivity. However, hog prices at the farm level actually declined by 3.3 percent. The difference between these price trends, 31.8 percent, represents the estimated decrease in the price of hogs at the farm gate attributable to productivity increases.

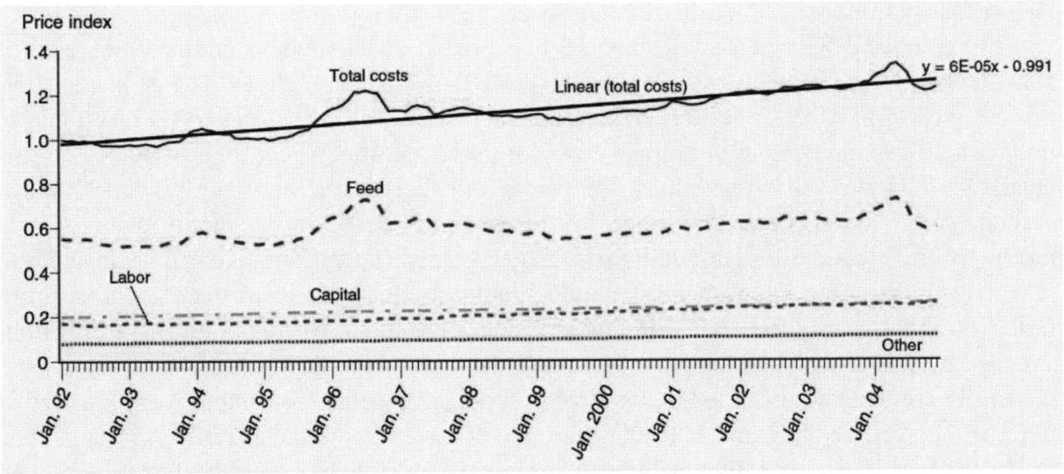

Sources: USDA, ERS using data from USDA, NASS, (farm wage rate); BLS (CPI; swine feeds, complete; farm machinery and equipment).

Figure 12. Historical pork prices—Gross farm value, wholesale value, and retail value, January 1992-December 2004

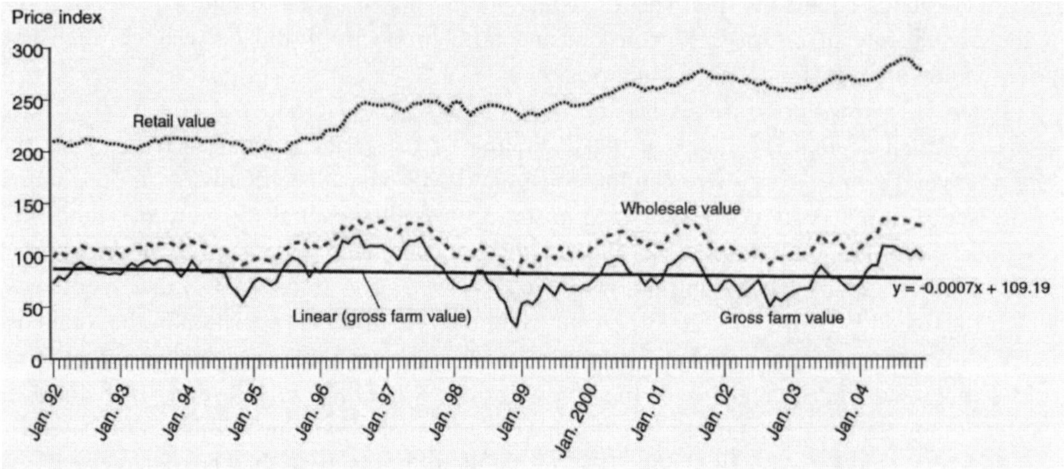

Sources: USDA, ERS using data from USDA, AMS, (wholesale and gross farm value); BLS (retail).

Figure 13. Feeder-to-finish hog production input price index, January 1992-December 2004

Productivity gains that lower finish-hog prices do not directly translate into lower retail prices. The cost of hogs represents less than a third of the total cost of retail pork (hogs must be slaughtered and processed, and pork must be transported and marketed).[11] Assuming finish hogs at the farm gate accounted for 29 percent of the total cost of pork sold at retail, farm-level productivity gains would lower retail pork prices by 9.2 percent (29 percent of 31.8 percent), compared with estimated prices without the productivity gains. Wholesale pork prices (at the packing house) show a very similar trend to the farm gate prices (see figure 13). In contrast, retail pork prices increased substantially over the same period. Factors that may have contributed to the rapid increase in retail prices include slower productivity growth in the retail sector, greater input price inflation for retailers, and increasing value added (see Hahn (2004) for more information about meat price spreads).

The estimated 31.8-percent decline in hog prices attributable to productivity gains is consistent with the estimates of TFP gains for feeder-to-finish operations. The estimated 109-percent growth in TFP for feeder-to-finish operations between 1992 and 2004 (see table 6) implies a 52-percent decline in average costs for these operations over this period.[12] The cost decline in feeder-to-finish operations is expected to exceed the overall hog price decline for several reasons. First, finish hog prices are determined partly by the supply of hogs from farrow-to-finish operations, in addition to feeder-tofinish operations. As shown in table 4, farrow-to-finish farms achieved substantially smaller productivity gains than feeder-to-finish farms, implying a smaller decline in costs. Second, the total cost of producing finish hogs supplied by feeder-to-finish operations includes the cost of producing feeder pigs. However, feeder-pig costs are not included in this study's productivity analysis (output was measured in terms of hundredweight gain). If feeder-pig operations realized smaller efficiency gains than feeder-to-finish operations, the total cost of supplying finish hogs would decline by less than 52 percent.

SUMMARY AND CONCLUSIONS

The U.S. swine industry has undergone significant changes in its size and structure during the past two decades that have coincided with substantial increases in hog farm productivity. In terms of structural change:

- The industry consolidated as the number of hog farms fell more than 70 percent between 1992 and 2004 while the hog inventory remained stable. Fewer and larger farms account for an increasing share of total output. The average size of U.S. hog operations grew from 945 head in 1992 to 2,589 head in 1998 and to 4,646 head in 2004. The share of the hog inventory on operations with 2,000 or more head increased from less than 30 percent to nearly 80 percent. Operations with 5,000 or more head accounted for more than 50 percent of the hog inventory in 2004.
- The organizational structure of U.S. hog farms changed considerably during the period as traditional farrow-to-finish operations gave way to large operations that specialize in a single phase of production. The share of market hogs produced by farrow-to-finish operations fell from 65 to 18 percent during 1992-2004. In contrast, the share produced by specialized hog-finishing operations increased from 22 to 77 percent.
- The expanded use of production contracts helped drive consolidation and specialization of hog production. The share of hogs delivered under production contracts grew from 5 percent of output in 1992 to 67 percent in 2004. Contracting operations are larger than independent operations and more likely to specialize in a single production phase.
- Farms with hogs became more specialized in the hog enterprise during 1992-2004 as the share of value of farm production from hog enterprises increased from 46 to 71 percent and the share of farm grain produced for hog feed fell from about half to below 20 percent. Changes have been particularly notable for specialized hog-finishing operations, where much of the production is under contract.
- The rapid growth of hog operations in the Southeast during 1992-98 slowed during subsequent years because of a moratorium on hog farm expansion in North Carolina, enacted in response to environmental concerns. In contrast, the size of hog operations increased faster in the traditional hog-producing States of the Midwest during 1998-2004.

Structural change has coincided with substantial efficiency gains for hog farms, particularly on the more specialized feeder-to-finish operations:

- Feeder-to-finish operations reduced the amount of feed used per unit of output by 4.7 percent annually between 1992 and 2004 and reduced the quantity of labor per unit by 13.8 percent annually.
- For feeder-to-finish producers, real average production costs per hundredweight of gain declined 4.7 percent annually between 1992 and 2004.

Productivity is a fundamental determinant of farm profitability and, hence, of farm growth and survival. This study used ARMS data to disaggregate changes in farm total factor productivity into changes in four constituent parts: technical change, technical efficiency, scale efficiency, and allocative efficiency. For feeder-to-finish farms:

- Total factor productivity increased annually by 6.4 percent from 1992 to 1998 and by 6.3 percent from 1998 to 2004. Most of these productivity gains were attributable to increases in the scale of production (scale efficiency) and technological change.
- The production technology displayed increasing returns to scale. Increases in farm size since 1992 explain almost half the total increase in farm productivity. Further increases in scale efficiency are likely limited for large farms, but scope for efficiency gains in the sector remains as smaller operations increase in scale or are replaced by larger operations.
- Technological advances helped drive almost half the total productivity increase since 1992.

Among findings related to a comparison of the two major hog-producing regions—the Southeast (mainly North Carolina) and the Heartland (mainly Iowa and Illinois):

- Regional trends in farm productivity mirrored trends in farm output: productivity increased more in the Southeast between 1992 and 1998 and increased more in the Heartland between 1998 and 2004.
- Average farm-size growth and the resulting improvements in scale efficiency explain most of the differences in productivity growth between the Heartland and Southeast since 1992. Farms in both regions had similar rates of technical advance over the study period. However, in the Southeast, rapid growth in average farm output during 1992-98 resulted in relatively large gains in scale efficiency. From 1998 to 2004, farms grew faster in the Heartland, leading to greater productivity growth in that region.

Recent ERS research associated use of production contracts with substantial increases in farm productivity. The estimates of productivity gains attributable to contracting suggest that these productivity advantages contributed to recent growth in contracting in the hog industry. The apparent continuing link between contracts and productivity, along with other benefits from contracts, suggests that the use of production contracts is likely to continue expanding. Because contract operations are generally larger operations, contracting may disproportionately benefit larger farms.

Cheaper food is one of the main potential benefits to society of greater hog farm productivity. The study found that productivity gains contributed to about a 30-percent reduction in the price of hogs at the farm gate, compared with prices estimated without the productivity gains. ERS did not estimate the extent to which the reduction in hog prices lowered retail pork prices. While productivity gains can benefit consumers in terms of lower food prices, structural changes that enable efficiency gains may also generate environmental concerns. Increases in the scale of production resulting in greater animal density may require operations to store manure in larger lagoons/pits—creating concentrated levels of odor, ammonia emissions, and the potential for larger manure spills. The concentration of hog manure makes it more costly to use as fertilizer as more land is needed and transportation costs to fields are greater. On the other hand, concentrating manure sources in fewer locations potentially affects fewer people. Additionally, greater concentration may make some manure treatment technologies feasible (e.g., energy from biowaste, or processing into concentrated fertilizer).

GLOSSARY

All-in/all-out housing commingles pigs of a similar age and weight and keeps them together as they move through each production phase. Marketing is done a room at a time, and rooms are washed and disinfected after each group of pigs leave to help decrease the spread of infectious diseases.

Commercial seed stock producers specialize in the production and sale of high-quality breeding hogs.

Contract production is an arrangement in which a pig owner (**contractor**) engages a producer (**grower**) to take custody of the pigs and care for them in the producer's facilities with other inputs often furnished by the pigs' owner. The producer is paid a fee for the service provided.

Farm Resource Regions portray the geographic distribution of U.S. farm production by identifying areas where similar types of farms intersect with areas of similar physiographic, soil, and climatic traits (USDA, ERS).

Hog operations are defined as farms that had a hog inventory of 25 head or more on the acres operated at any time during survey years 1992, 1998, and 2004. Hog operations include independent hog producers and growers who produced hogs under contract.

Hundredweight gain equals hundredweight of hogs sold or removed under contract less hundredweight of hogs purchased or placed under contract, plus hundredweight of inventory change each year, expressed as:

$$CWTGAIN = (CWTSR - CWTPP) + (CWTEINV - CWTBINV)$$

where CWTGAIN is hundredweight gain, CWTSR is hundredweight of sales and contract removals, CWTPP is hundredweight of purchases and contract placements, CWTEINV is hundredweight of inventory on December 31, and CWTBINV is hundredweight of inventory on January 1.

Operating costs are the costs for purchased input items that are consumed during one production period. These include feed; feeder pigs; veterinary and medical services; marketing; custom services and supplies; fuel, lubrication, and electricity; repairs; hired labor; and operating capital.

Ownership costs are the costs associated with the ownership of depreciable assets, such as farm tractors and hog-production facilities. These include depreciation, interest, property taxes, and insurance.

Phase feeding feeds hogs or pigs diets of varying protein and energy content at different stages, or phases, of their life to more closely match the diet with their changing nutritional requirements.

Phase of production refers to one of four commonly used categories that describe stages in the hog-production process: (1) breeding and gestation— the breeding of females and their maintenance during the gestation period; (2) farrowing—the birth of baby pigs until weaning; (3) nursery—the care of pigs from immediately after weaning until about 30-80 pounds, and; (4) growing/finishing—the feeding of hogs from 30-80 pounds to the slaughter weight of 225-300 pounds.

Terminal crossbreeding programs concentrate on using all possible heterosis of the breeds and thus capitalize on breed strengths. These programs use two-, three-, or four-breed first-cross females that excel in maternal traits bred to boars from breeds that are superior for growth and carcass traits. All the progeny from these matings are marketed and not kept for replacement gilts.

Total economic costs are the full ownership costs (cash and noncash) for being engaged in the enterprise. These include both operating and ownership costs, plus opportunity costs for unpaid labor and land, and costs for general farm overhead items.

Type of hog producer is a classification that defines the hog operation according to the phases of production conducted on the operation and the type of product produced. Some operations in each survey could not be classified using the following criteria:

> **Farrow-to-finish** operations are those on which pigs are farrowed and then finished to a slaughter weight of 225-300 pounds. Using the survey data, these operations were defined as farms on which more than 75 percent of pigs came from onfarm farrowings and more than 75 percent of the value of hogs and pigs left the operation through market hog sales or contract removals.
>
> **Farrow-to-feeder** operations are those on which pigs are farrowed and then sold or removed under contract at or after weaning at a weight of about 30-80 pounds. Using the survey data, these operations were defined as farms on which more than 75 percent of pigs came from onfarm farrowings and more than 75 percent of the value of hogs and pigs left through feeder pig sales or contract removals.
>
> **Farrow-to-wean** operations are those on which pigs are farrowed and then sold or removed under contract after an early weaning at a weight of about 10-20 pounds. Using the survey data, these operations were defined as farms on which more than 75 percent of pigs came from onfarm farrowings and more than 75 percent of the value of hogs and pigs left through weanling sales or contract removals.

Feeder-to-finish operations are those on which feeder pigs are obtained from outside the operation, either purchased or placed under contract, and then finished to a slaughter weight of 225-300 pounds. Using the survey data, these operations were defined as farms on which more than 75 percent of pigs came from feeder pig purchases or contract placements and more than 75 percent of the value of hogs and pigs left through market hog sales or contract removals.

Wean-to-feeder operations are those on which weanlings (10-20 pounds) are obtained from outside the operation, either purchased or placed under contract, and then fed to a feeder pig weight of about 30-80 pounds. Using the survey data, these operations were defined as farms on which more than 75 percent of pigs came from weanlings purchased or placed under contract and more than 75 percent of the value of hogs and pigs left through feeder pig sales or contract removals.

REFERENCES

Ahearn, M., Yee, J., Ball, E. & Nehring, R. (1998). *Agricultural Productivity in the United States, Agriculture Information Bulletin No. 740*, U.S. Department of Agriculture, Economic Research Service, January.

Battese, G. E., & Coelli, T. J. (1992). "Frontier Production Functions, Technical Efficiency and Panel Data: With Application to Paddy Farmers in India," *Journal of Productivity Analysis, 3*, 153-169.

Boessen, C., Lawrence, J. D. & Grimes, G. (2004). *Production and Marketing Characteristics of U.S. Pork Producers - 2003*, Working Paper No. AEWP 2004-04, University of Missouri, Department of Agricultural Economics, July.

Bureau of Economic Analysis. *Gross Domestic Product by Industry Accounts*, http://www.bea..gov/bea/industry/gpotables, accessed May 22, 2006.

Coelli, T., Estache, A., Perelman, S. & Trujillo, L. (2003). *A Primer on Efficiency Measurement for Utilities and Transport Regulators*, The World Bank.

Coelli, T. J., Rao, D., O'Donnell, C. J. & Battese, G. E. (2005). *An Introduction to Efficiency and Productivity Analysis, 2nd Edition*, Springer.

Hahn, W. F. *Beef and Pork Values and Price Spreads Explained*, Outlook Report No. LDP-M-118-01, U.S. Department of Agriculture, Economic Research Service, May 2004, www.ers.usda.gov/publications/ldp/apr04/ldpm11801/

Haley, M. M. *Market Integration in North American Hog Industries*, Outlook Report No. LDP-M-125-01, U.S. Department of Agriculture, Economic Research Service, November 2004.

Key, N. & McBride, W. "Production Contracts and Productivity in the U.S. Hog Sector," *American Journal of Agricultural Economics*, February 2003, *Vol. 85*, No. 1, pp. 121-133.

Key, N. & McBride, W. (2007) *"Do Contracts Raise Hog Farm Productivity? An Instrumental Variables Approach,"* selected paper at the annual meeting of the American Agricultural Economics Association, Portland, OR, July.

McBride, W. D. Hog, U. S. (1995). *Production Costs and Returns, 1992: An Economic Basebook*, Agricultural Economics Report No. 724, U.S. Department of Agriculture, Economic Research Service, November.

McBride, W. D. & Key, N. (2003). *Economic and Structural Relationships in U.S. Hog Production*, Agricultural Economics Report No. 818, U.S. Department of Agriculture, Economic Research Service, February.

North Carolina General Assembly. The Clean Water Responsibility and Environmentally Sound Policy Act, 1997, http://www.ncga.state.nc.us/gascripts/BillLookUp/BillLookUp.pl?Session=1997&BillID= H515, accessed October 5, 2006.

Orea, L. (2002). "Parametric Decomposition of a Generalized Malmquist Productivity Index," *Journal of Productivity Analysis 18*, 5-22.

Successful Farming. *Pork Powerhouses 2005: 20 Largest Pork Producers in the United States*, October 2005.

U.S. Department of Agriculture, Animal and Plant Health Inspection Service, Veterinary Service (USDA, APHIS). *Swine 2000-Part IV: Changes in the U.S. Pork Industry*, 1990-2000, April 2005.

U.S. Department of Agriculture, Economic Research Service (USDA, ERS). ERS Farm Resource Regions, January 2006, www.ers.usda.gov/Briefing/ARMS/resourceregions/resourceregions.htm# new, accessed October 5, 2006.

U.S. Department of Agriculture, Economic Research Service (USDA, ERS). "*Meat Price Spreads*," www.ers.usda.gov/Data/meatpricespreads/, accessed April 9, 2007.

U.S. Department of Agriculture, Grain Inspection, Packers, and Stockyard Administration (USDA, GIPSA). *GIPSA Livestock and Meat Marketing Study Volume 1: Executive Summary and Overview,* January 2007.

U.S. Department of Agriculture, National Agricultural Statistics Service (USDA, NASS). *Farms, Land in Farms, and Livestock Operations: Annual Summary,* January 2005.

U.S. Department of Agriculture, National Agricultural Statistics Service (USDA, NASS). *Agricultural Prices: Annual Summary*, July 1996-2005.

U.S. Department of Agriculture, National Agricultural Statistics Service (USDA, NASS). *Hogs and Pigs,* December 1995-99.

U.S. Department of Agriculture, National Agricultural Statistics Service (USDA, NASS). *Meat Animals Production, Disposition, and Income: 2004 Summary,* April 2005.

Van Arsdall, R.N., and K.E. Nelson. U.S. Hog Industry, Agricultural Economics Report No. 511, U.S. Department of Agriculture, Economic Research Service, June 1984.

APPENDIX

Decomposition of total factor productivity

This study uses a stochastic frontier analysis to decompose TFP growth into four components: (1) technical change, which is the increase in the maximum output that can be produced from a given level of inputs (a shift in the production frontier); (2) technical efficiency change, which is the change in a firm's ability to achieve maximum output given its set of inputs (how close it is to the production frontier); (3) scale efficiency change, which is the change in the degree to which a firm is optimizing the scale of its operations; and (4) allocative efficiency change, which is the change in a firm's ability to select a level of inputs so as to ensure that the input price ratios equal the ratios of the corresponding marginal products.[13]

Orea (2002) shows that if firm i's technology in time t can be represented by the translog output-oriented distance function Do(qit , xit ,t) where q is output, xit, a K-dimensional input vector with elements (xit1 , . ..xitk ...xitK), then the logarithm of a generalized output-oriented Malmquist productivity index ln Mo can be decomposed into changes in technical efficiency (EC), technical change (TC), and scale efficiency change (SC), between time periods r and s:

$$\ln M_{Oi} = EC_i^{rs} + TC_i^{rs} + SC_i^{rs} \tag{1}$$

where

$$EC_i^{rs} = \ln D_O(q_{is}, x_{is}, s) - \ln D_O(q_{ir}, x_{ir}, r) \tag{2}$$

$$TC_i^{rs} = -\frac{1}{2}\left[\frac{\partial \ln D_O(q_{is}, x_{is}, s)}{\partial t} + \frac{\partial \ln D_O(q_{ir}, x_{ir}, r)}{\partial t}\right] \tag{3}$$

$$SC_i^{rs} = \frac{1}{2}\sum_{k=1}^{K}\left[\frac{\varepsilon_{is}-1}{\varepsilon_{is}}\varepsilon_{isk} + \frac{\varepsilon_{ir}-1}{\varepsilon_{ir}}\varepsilon_{irk}\right]\cdot \ln\left(\frac{x_{isk}}{x_{irk}}\right) \tag{4}$$

where $t = (r,s)$, $\varepsilon_{it} = \sum_{k=1}^{K} \varepsilon_{itk}$ is the scale elasticity such that $\varepsilon_{itk} = \partial \ln D_O(q_{it}, x_{it}, t) / \partial \ln x_{itk}$.

With one output, a translog distance function can be defined:

$$\ln D_O(q_{it}, x_{it}, t) = \ln q_{it} - f(\beta, x_{it}, t) - v_{it} \tag{5}$$

where âis a vector of parameters, *vit* is a normally distributed random error with mean zero and:

$$f(\beta, x_{it}, t) = \beta_0 + \sum_{k=1}^{K}\beta_k \ln x_{itk} + \frac{1}{2}\sum_{k=1}^{K}\sum_{j=1}^{K}\beta_{kj}\ln x_{itk}\ln x_{itj} + \sum_{k=1}^{K}\beta_{tk} t \ln x_{itk} + \beta_t t + \frac{1}{2}\beta_{tt} t^2 . \tag{6}$$

To account for technical inefficiency, we estimate a stochastic production function model of the form:

$$\ln q_{it} = f(\beta, x_{it}, t) + v_{it} - u_{it} \quad (7)$$

where u_{it}, a nonnegative random variable associated with technical inefficiency, is drawn from a truncated normal distribution (Battese and Coelli, 1992). An output-oriented measure of technical efficiency is the ratio of observed output to the corresponding stochastic frontier output:

$$TE_{it} = \frac{q_{it}}{\exp(f(\beta, x_{it}, t) + v_{it})} = \frac{\exp(f(\beta, x_{it}, t) + v_{it} - u_{it})}{\exp(f(\beta, x_{it}, t) + v_{it})} = \exp(-u_{it}) \quad (8)$$

Using (7), it can be shown that the technical efficiency factor (8) equals the distance function (5):

$$\exp(-u_{it}) = \exp(\ln q_{it} - f(\beta, x_{it}, t) - v_{it}) = D_0(q_{it}, x_{it}, t). \quad (9)$$

The technical efficiency measure (8) can be estimated conditional on $e_{it} = v_{it} - u_{it}$. It follows from (2) and (8) that the efficiency change can be

$$EC_i^{rs} = E(-u_{is} \mid e_{is}) - E(-u_{ir} \mid e_{ir}) \quad (10)$$

or

$$\exp(EC_i^{rs}) = E(\exp(-u_{is} \mid e_{is})) / E(\exp(-u_{ir} \mid e_{ir})), \quad (11)$$

where the numerator and denominator in (11) are the estimated technical efficiency scores in periods s and r, respectively, which have values between zero and one.

Using (3), (5), and (6), the technical change index can be derived:

$$TC_i^{rs} = \frac{1}{2}\left[\sum_{k=1}^{K} \beta_{tk} \ln x_{isk} + \sum_{k=1}^{K} \beta_{tk} \ln x_{irk} + 2\beta_t + 2\beta_{tt}(r+s)\right]. \quad (12)$$

From (6), the scale elasticity is:

$$\varepsilon_{it} = \sum_{k=1}^{K} \varepsilon_{itk}, \text{ where } \varepsilon_{itk} = \beta_k + \frac{1}{2}\sum_{j=1}^{K} \beta_{kj} \ln x_{itj} + \beta_{tk} t \quad (13)$$

This can be used to compute the scale efficiency change index (4).

To estimate allocative efficiency change, ERS compared the Malmquist TFP index (1) to the logarithm of the Tornqvist TFP change index (with one output):

$$\ln TFP_i^{rs} = \ln q_{is} - \ln q_{ir} - \frac{1}{2}\sum_{k=1}^{K}\left[(\sigma_{isk} + \sigma_{irk})\cdot(\ln x_{isk} - \ln x_{irk})\right] \tag{14}$$

where óitk are the input cost shares. Any difference between the Tornqvist TFP change calculated in (14) and the Malmquist TFP index calculated in (1) must be due to allocative efficiency change. Hence, it can be shown that the allocative efficiency change is:

$$AC_i^{rs} = \frac{1}{2}\sum_{k=1}^{K}\left[\left(\left(\frac{\varepsilon_{isk}}{\varepsilon_{sk}} - \sigma_{isk}\right) + \left(\frac{\varepsilon_{irk}}{\varepsilon_{rk}} - \sigma_{irk}\right)\right)\cdot(\ln x_{isk} - \ln x_{irk})\right] \tag{15}$$

Appendix table 1. Stochastic production function parameter estimates

Parameter		Coefficient	Standard error	t-statistic
β_0	constant	0.377	0.0385	9.8
β_1	feed	0.473	0.0214	22.2
β_2	labor	0.045	0.0119	3.8
β_3	capital	0.319	0.0258	12.4
β_4	other inputs	0.280	0.01 93	14.5
$\beta 11$		0.101	0.0323	3.1
$\beta 22$		-0.028	0.0148	-1.9
$\beta 33$		0.092	0.0609	1.5
$\beta 44$		0.081	0.0337	2.4
$\beta 12$		-0.0055	0.01 88	-0.3
$\beta 13$		-0.0791	0.0383	-2.1
$\beta 14$		-0.0738	0.0268	-2.8
$\beta 23$		0.0060	0.0207	0.3
$\beta 24$		-0.0183	0.0174	-1.1
$\beta 34$		0.0226	0.0366	0.6
β_t	time	0.0619	0.0034	18.2
βtt	time-squared	0.0046	0.0017	2.7
$\beta t1$		-0.0257	0.0045	-5.7
$\beta t2$		0.0012	0.0029	0.4
$\beta t3$		0.0065	0.0058	1.1
$\beta t4$		0.0212	0.0043	4.9
$\sigma 2$	(=ó2 $_v$ + ó^2_u)	0.355	0.0300	11.8
γ	(=ó2 $_v$ + ó^2_u)	0.725	0.0536	13.5
Observations		1,181		

Source: USDA, ERS using data from the 1992 Farm Costs and Returns Survey and USDA's 1998 and 2004 Agricultural Resource Management Surveys.

In the analysis, output is defined as "hog weight gain"—the weight added to purchased/placed hogs and existing hog inventory in the calendar year prior to the year of the survey. Feed is defined as the total weight of feed applied. The labor input is a Tornqvist quantity index comprised of paid labor and unpaid farm household labor using the labor expenditure shares for paid and unpaid labor as weights. Capital is the "capital recovery cost"—the estimated cost of replacing the existing capital equipment (barns, feeding equipment, etc.). "Other inputs" is defined as expenditures on veterinary services, bedding, marketing, custom work, energy, and repairs. Labor wages are deflated using the Bureau of Labor Statistics (BLS) Blue Collar Total Compensation index; feed prices are deflated using a weighted average of the BLS corn and soybean Producer Price Index (PPI); Capital is deflated using the BLS farm machinery PPI, and other inputs are deflated using the CPI. In the estimation, ERS rescaled all logged values of the variables as deviations from the sample mean to facilitate interpretation of the coefficients.

The appendix table presents the estimated coefficients of the stochastic production function (6). Because the variables are expressed as deviations from their means, the first-order parameters of the translog function can be directly interpreted as estimates of production elasticities evaluated at the sample means. The production elasticities with respect to feed, capital, and other inputs have plausible values and are statistically significant.

End Notes

[1] Hog inventory is the number of hogs and pigs on an operation at a point in time. Often, this is the beginning or ending of a year but may refer to the maximum number of hogs on the operation at any time during a year. In contrast, hog production refers to the flow of hogs on and off the operation during a year.

[2] Specialized farrowing and weanling rearing operations provide feeder pigs to feeder-to-finish operations that finish hogs to a market weight. Hogfinishing operations may also obtain feeder pigs from other countries. The number of hogs imported for finishing from Canada has grown significantly in recent years (Haley, 2004).

[3] Each head produced represents approximately a 2-cwt gain (250 pounds for a typical finished market hog minus 50 pounds for a typical feeder pig). Therefore, ignoring losses due to animal mortality, a farm with an output of 6,000 cwt gain removes 3,000 head per year. Assuming three hog cycles per year, annual production of 6,000 cwt implies an operation has an inventory of 1,000 head.

[4] For example, if a farm has an estimated scale elasticity of 1.25, then a 1-percent increase in all inputs would result in a 1.25-percent increase in output. Scale elasticity above 1 implies increasing returns to scale. Scale elasticity equal to 1 implies constant returns to scale—that is, there is no productivity gain (or loss) from increasing the scale of production. Scale elasticity below 1 implies decreasing returns to scale—efficiency would actually decline if production were increased.

[5] Increasing returns to scale are usually associated with declining per unit costs as output increases. However, declining per unit costs do not always imply increasing returns to scale. Lower per unit costs for large producers may also result if technical efficiency or allocative efficiency increases with scale, or if input costs decrease with scale. In addition, it is not necessarily the case that all inputs increased proportionately as scale increased.

[6] ARMS did not survey enough farrow-to-wean or wean-to-feeder pig operations in 1992 and 1998 to measure industry-wide efficiency gains from increased specialization.

[7] Feeder-to-finish operations are those on which feeder pigs (weighing 30-80 pounds) are purchased/placed, finished, and then sold/removed for slaughter (weighing 225-300 pounds).

[8] The procedure for estimating the stochastic production function allows some parameters to vary over time, which permits scale elasticity and optimal scale to vary also. See the definition of scale elasticity in the appendix for more details.

[9] For full text of the bill, see: http://ssl.csg.org/dockets/99bscbills/ 2499b01nchb515cleanswine.html

[10] For each region, the disaggregation used the estimated parameters for all farms and the input levels corresponding to the farms in that region. In other words, the production technology was assumed to be the same across regions, but the input mix varied according to the sample.

[11] The share of the farm value of pork (net byproducts) in the retail value fluctuated between 26 and 33 percent between 2003 and 2006, and averaged 28.9 percent (see www.ers.usda.gov/data/ meatpricespreads/data/pork.xls).

[12] A TFP index is proportional to the inverse of an average cost index. From table 3, TFP was estimated to increase by 109.1 percent between 1992 and 2004, implying that a TFP index equal to 1.0 in 1992 would increase to 2.091 in 2004. The inverse of this index (the average cost), would fall from 1.0 to 0.478—a 52.2-percent decline.

[13] The derivations in this appendix are based primarily on Orea (2002); Coelli et al. (2005), pp. 289-302; and Coelli et al. (2003), pp. 25-66.

In: Transformation of U.S. Animal Agriculture
Editor: Justin M. Daigle
ISBN: 978-1-60876-938-4

Chapter 4

PROFITS, COSTS, AND THE CHANGING STRUCTURE OF DAIRY FARMING

James M. MacDonald, Erik J. O'Donoghue, William D. McBride, Richard F. Nehring, Carmen L. Sandretto and Roberto Mosheim
U.S. Dept. of Agriculture (USDA)

ABSTRACT

U.S. dairy production is consolidating into fewer but larger farms. This report uses data from several USDA surveys to detail that consolidation and to analyze the financial drivers of consolidation. Specifically, larger farms realize lower production costs. Although small dairy farms realize higher revenue per hundredweight of milk sold, the cost advantages of larger size allow large farms to be profitable, on average, even while most small farms are unable to earn enough to replace their capital. Further survey evidence, as well as the financial data, suggest that consolidation is likely to continue.

Keywords. Dairy farming, economies of scale, economies of size, dairy farm structure, milk costs

ACKNOWLEDGMENTS

The authors thank Don Blayney, Noel Gollehon, Marc Ribaudo, and Neil Conklin of the Economic Research Service, Howard McDowell and Cliff Carman of the Agricultural Marketing Service, Jeff Gillespie of Louisiana State University, Leslie Butler of the University of California at Davis, and an anonymous external reviewer for comments and advice. They also thank Dale Simms for excellent editing support, and Wynnice Pointer-Napper for a timely and effective design and layout of the report.

SUMMARY

Dairy farming in the United States is undergoing dramatic changes, driven by both supply and demand factors. Consumption is shifting from fluid milk, generally produced for local markets, toward manufactured products, such as cheese, and dairy-based ingredients produced for national and global markets. Innovations in breeding and feeding systems have led to large increases in the amount of milk that a cow produces. Milk production is shifting toward Western States such as California, Idaho, and New Mexico, and to much larger farms. The number of dairy farms with fewer than 200 cows is shrinking, while the number of very large operations, with 2,000 or more cows, doubled between 2000 and 2006.

What Is the Issue?

Large dairy farms first emerged in the Western States, but are now appearing in traditional dairy States as well. This report documents shifts in the location and size of dairy farms and takes a look at what those changes may mean. If the shift in farm size reflects economies of scale in dairy production—that is, lower costs on larger farms—then increasing farm size also enables milk to be produced with fewer resources, thereby reducing prices to consumers. However, the shifts also concentrate animal wastes from manure onto a much smaller land base and may exacerbate pollution associated with concentrated livestock production.

What Did the Study Find?

Large dairy enterprises generate returns that, on average, well exceed their full costs. At the same time, smaller dairy farms mostly incur economic losses—the value of their production does not exceed full costs, including the costs of capital and time committed by their owners. Large farms incur much lower costs, on average, than smaller farms, and these advantages accrue across a wide range of sizes. Costs per hundredweight of milk produced fall by nearly half as herd size increases from fewer than 50 head to 500 head, and continue to fall, but less sharply, at even larger herd sizes.

Dairy investment decisions are consistent with the financial evidence. Farms with fewer than 200 cows accounted for over two-thirds of the nationwide inventory of cows in 1992. By 2006, their share of the nationwide inventory had dropped to 38 percent. Meanwhile, farms with at least 1,000 head of dairy cows are growing more prevalent. They accounted for less than 10 percent of inventory in 1992 but more than a third by 2006. Structural shifts are evident among the largest farms, too. During the 1990s, farms with 1,000–3,000 head were adding the most capacity, but capacity additions have since shifted to even larger farms, with 3,000–10,000 head.

Some small dairy farms are profitable, and others continue to earn enough to remain in operation. As a result, structural change is likely for the foreseeable future, with a continuing decline, rather than a sudden disappearance, of small and midsize dairy operations. The ongoing structural changes will continue to place downward pressure on milk prices.

Excess nutrient applications, which arise from animal manure and can cause water and air pollution, appear to be intensified on larger operations. But their production cost advantages still outweigh the likely additional costs of manure treatment and removal, and it is unlikely that manure management regulations will reverse the ongoing patterns of structural change.

How Was the Study Conducted?

Confidential farm-level records from successive censuses of agriculture (1992, 1997, and 2002) were used to depict changes in the location and size distribution of dairy farms. More aggregated public information on the size distribution of dairy farms is drawn from annual dairy surveys carried out by USDA's National Agricultural Statistics Service (NASS).

Additional farm-level data come from the annual Agricultural Resource Management Survey (ARMS), administered jointly by the Economic Research Service (ERS) and NASS. Dairy farms in major milk-production States were targeted with commodity-specific ARMS versions covering operations during 2000 and 2005. These surveys provide detailed information for analyses of costs, manure management practices, and operator expectations for survival. Data from successive years of version 1 of the ARMS are used to develop measures of potential excess nutrient production on dairy operations.

This study focuses on conventional dairy production, and does not assess costs and farm sizes among organic dairy operations, a rapidly growing but still small segment of the industry. The 2005 ARMS dairy version contains comprehensive information on a sample of organic producers, and other research projects are analyzing those data.

INTRODUCTION

Dairy farming in the United States is undergoing dramatic changes, driven by both supply and demand factors. Consumption is shifting from fluid milk, generally produced for local markets, toward manufactured products, such as cheese, and dairy-based ingredients produced for national and global markets. Innovations in breeding and feeding systems have led to large increases in the amount of milk that a cow produces. The location of milk production is shifting toward Western States such as California, Idaho and New Mexico. Finally, production is shifting to much larger farms. The number of dairy farms with fewer than 200 cows is shrinking rapidly while very large operations, with 1,000 to 30,000 cows on one site, account for rapidly growing shares of production. Large dairy farms first emerged in the Western States, but are now appearing in traditional dairy States as well.

Earlier Economic Research Service (ERS) and USDA studies document the broad patterns of structural change in the dairy sector (Blayney, 2002; Miller and Blayney, 2006; U.S. Department of Agriculture, 2004). This report focuses on issues surrounding the growing size of dairy operations and the closely linked factor of location. The increasing share of larger farms suggests that they have cost advantages over small operations, but the size of such advantages, and the range of herd sizes over which they apply, is uncertain. Knowledge of each dimension of scale economies is crucial to understanding the structural changes in the industry. Finally, the report evaluates the links among dairy farm consolidation, concentration of cow manure, and manure management strategies and regulations.

This report focuses primarily on conventional (nonorganic) dairy farms. Although our analysis of farm size and locations covers all dairy farms, the cost structure of organic farms is sufficiently different from conventional operations as to require a separate cost analysis.

CHANGES IN THE SIZE AND LOCATION OF U.S. DAIRY FARMS

The emergence of large dairy farms, and the continued shift of production toward such farms, is the principal focus of our analysis. We also touch on geographic shifts in production, a closely intertwined component of structural change.

Between 1970 and 2006, the number of farms with dairy cows fell steadily and sharply, from 648,000 operations in 1970 to 75,000 in 2006, or 88 percent (Figure 1). Total dairy cows fell from 12 million in 1970 to 9.1 million in 2006, so the average herd size rose from just 19 cows per farm in 1970 to 120 cows in 2006.[1] Moreover, because milk production per cow doubled between 1970 and 2006 (from 9,751 to 19,951 pounds per year), total milk production rose, and average milk production per farm increased twelvefold.

These changes reflect a trend toward greater specialization as well as greater size. However, like much of agriculture, dairy farms come in a wide range of sizes. The largest U.S. dairy farms have over 15,000 cows, though farms with 1,000–5,000 cows are more common. Large dairy farms account for most inventory and production in Western States, and a growing share of production elsewhere.

The smallest class of dairy farms (fewer than 30 cows) still accounted for nearly 30 percent of all operations with milk cows in 2006, but had only 2 percent of all cows and provided just over 1 percent of total dairy production (table 1). Such farms, which frequently combine a very small dairy enterprise with other commodity enterprises or with off-farm work, are disappearing rapidly.

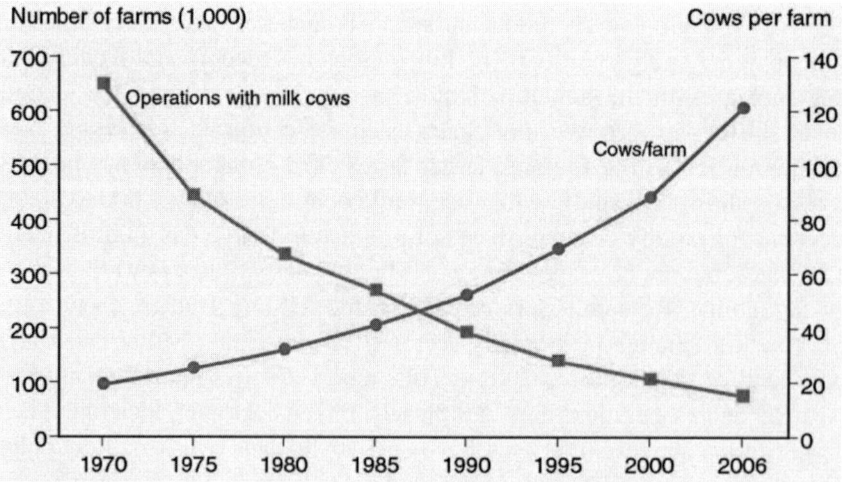

Source: USDA, NASS.

Figure 1. The number of dairy farms is declining, while average size is growing

Table 1. Changes in the size structure of U.S. dairy farms, 2000-2006

Herd size No. Head	Number of operations		% change	Percent of inventory		Percent of production	
	2000	2006		2000	2006	2000	2006
1-29	30,810	21,280	-31.0	2.9	1.9	1.8	1.2
30-49	22,110	14,145	-36.0	9.1	6.0	7.7	4.9
50-99	31,360	22,215	-29.2	22.0	16.3	19.4	14.3
100-199	12,865	9,780	-24.0	18.0	14.1	17.3	13.0
200-499	5,350	4,577	-14.4	16.7	15.0	18.0	15.0
500-999	1,700	1,700	0	12.0	12.6	13.7	14.3
1,000-1,999	695	870	+25.2	10.1	12.5	11.6	13.9
2,000+	280	573	+104.6	9.2	21.6	10.5	23.4
Total	105,170	75,140	-25.5	100.0	100.0	100.0	100.0

Source: USDA, NASS Milk Production, Feb. issue (through 2004); USDA, NASS Farms, Land in *Farms and Livestock Operations* (after 2004). Herd size refers to all dairy cows on an enterprise, including dry cows but excluding calves, heifers, and bulls.

Changes in the location of milk production are closely intertwined with changes in farm size. In table 2, production and structure indicators are reported for each of the 16 largest dairy States, which together account for 83 percent of U.S. milk production. For each State, we report milk production and the share of a State's production in small (fewer than 100 cows) and large (500 or more cows) farms.[2] Production data are for 1994, 2000, and 2006 (the most recent available) and structure indicators are for 2000 and 2006.

The next three size classes (30–200 cows) tend to specialize in dairying. These classes are also in sharp decline, with farm numbers falling by 30 percent between 2000 and 2006 (table 1). Production is shifting to farms with at least 500 cows, with the most striking changes occurring in dairies with at least 2,000 milk cows. The number of farms in this largest size class more than doubled between 2000 and 2006, as did its shares of cow inventory and total milk production.

Large and small dairy farms are organized in fundamentally different ways (Short, 2004; Sumner and Wolf, 2002). Large farms usually purchase significant amounts of feed and contract with other operations to raise their heifers offsite. Small farms grow more of their own feed and raise their heifers onsite. Large operations tend to confine their milk cows in large barns or in drylot feedyards, while small operations may graze their cows on pasture. Most labor on small dairy farms is provided by the operator and the operator' s family, whereas large farms rely extensively on hired labor (although they are usually family-owned and operated).

Large farms dominate in California, the Nation's largest milk-producing State. Farms with at least 500 cows accounted for 88 percent of California's production in 2006, and production there grew by more than half between 1994 and 2006, as the State's share of national production rose from 16 to 21 percent (table 2). Other States in the West and Southwest show similar patterns—substantial growth in production and a concentration in large dairy farms.

Table 2. Milk production and farm structure in major dairy States

State	Production			Herd size			
				<100 head		>499 head	
	1994	2000	2006	2000	2006	2000	2006
	(Billion pounds)			(Percent of State production)			
Northeast	24.0	25.5	25.3	46.5	38.9	10.4	21.3
NY	11.4	11.9	12.0	34.0	28.5	16.0	31.0
PA	10.2	10.9	10.7	63.0	53.0	3.0	10.0
VT	2.4	2.7	2.6	35.0	26.0	16.0	29.0
E. Corn Belt	12.3	12.8	15.3	39.9	28.4	13.1	31.2
IN	2.3	2.6	3.3	51.0	29.0	10.0	43.0
MI	5.5	5.9	7.1	28.0	18.0	20.0	39.0
OH	4.5	4.3	4.9	49.5	36.0	5.5	23.0
Upper Midwest	31.7	31.0	31.8	56.7	48.2	8.9	15.6
MN	9.3	8.8	8.4	59.5	47.5	8.5	17.5
WI	22.4	22.2	23.4	56.0	45.0	9.0	19.0
Southwest	11.6	13.6	17.8	2.8	1.6	78.2	87.3
NM	3.3	5.6	7.6	0.2	0.2	98.0	98.0
AZ	2.1	2.9	3.7	0.4	0.5	95.0	98.0
TX	6.2	5.1	7.1	7.0	2.0	47.0	78.0
West	37.3	50.1	59.9	1.8	1.3	73.4	84.2
CA	25.0	33.3	38.8	0.6	0.5	78.0	88.0
CO	1.6	1.9	2.5	3.0	1.9	63.0	83.0
ID	3.8	7.8	10.9	4.5	2.0	74.0	89.0
OR	1.7	1.6	2.2	8.0	6.0	39.0	54.0
WA	5.2	5.5	5.5	3.0	2.4	58.0	70.0
16 major States	116.9	133.0	150.1	26.9	20.4	41.0	54.0
US	153.6	167.6	181.8	28.9	20.4	35.8	51.6

Source: USDA, NASS *Milk Production*, monthly issues (through 2004); USDA, NASS F*arms, Land in Farms and Livestock Operations* (after 2004).

Milk production in traditional dairy States in the Northeast, Eastern Corn Belt, and Upper Midwest comes more from small dairies than from large. Although the three regions together maintained stable milk production volumes in 1994–2006, their share of national production fell by 4.5 percentage points.

But structural change is not simply a matter of regional differences. Large farms' share of milk production is increasing in every major dairy State. According to newspaper reports, over 40 large farms, each with 1,000- 5,000 cows, were built in Michigan, Ohio, and Indiana between 1998 and 2006. Farms with upwards of 1,000 head are also appearing in other traditional dairy States in the East and Midwest, either through the expansion of longstanding family operations or through new construction with investor financing.[3]

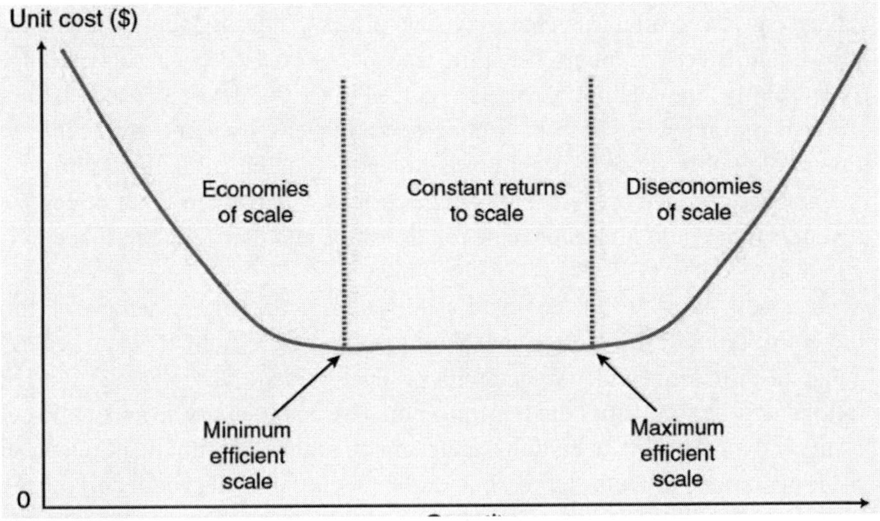

Source: USDA, ERS.

Figure 2. Visualizing scale economy concepts

SCALE ECONOMIES AND STRUCTURE IN DAIRY FARMING: BACKGROUND

Strong structural changes—specifically, the ongoing shift of production to larger operations—suggest that there may be significant economies of scale in dairy production, in the form of cost advantages accruing to increased herd sizes. This report assesses the sources, magnitude, and extent of scale economies, and traces their impact on the industry. However, there are several elements to the link between scale economies and farm structure, and we must first describe those elements.

A longrun average cost curve is depicted in figure 2.[4] The figure consists of three regions. At low levels of output, average costs decline with increases in output. The figure also displays a range of constant average costs, in which average costs do not vary with increases in output, and a range of diseconomies of scale (rises in average cost as production increases).[5]

Five elements of figure 2 are important for evaluating scale in an industry:

- The level of output at which scale economies are just exhausted (and at which constant returns set in) is called the firm's *minimum efficient* scale.
- The *cost penalty from small scale*—how much higher are the costs of small firms that are unable to realize minimum efficient scale?
- The output level at which diseconomies set in is called the *maximum efficient scale*—the largest firm size that can be achieved while still realizing all scale economies. Diseconomies are clearly important in agriculture, since even very large farms are still fairly small businesses.

- Cost curves are *efficiency frontiers*—they reflect the minimum costs that a firm can achieve, given available technology and prices paid for inputs. In practice, actual costs could exceed frontier costs (and thus be *inefficient*) because some inputs are in fixed supply and cannot easily be adjusted to the level needed to achieve the efficiency frontier, because of a poor operating environment (reflecting weather or topography) or because the operator is less effective than other operators.
- A cost curve reflects a given set of input prices. Changes in input prices would shift the curve, but could also alter scale relationships and therefore the shape of the curve.

When we assess how cost-scale relationships affect the size structure of farms, it is important to consider all of these elements. Minimum and maximum efficient scales drive the potential range of farm sizes and, coupled with product demand, largely determine how many farm operations will be in business in the long run. The cost penalty from small scale affects the likely survival of smaller operations that cannot realize minimum efficient scale. The efficiency of operations affects survival and the actual industrywide cost changes from structural change. Finally, relative price changes could alter the existing pattern of scale advantages. Increases in prices paid for hired labor, purchased feed, or manure transportation would, all else being equal, raise costs more for large dairy farms than for small since they use those inputs more intensively.

Explicit and Implicit Costs in Dairy Production

In assessing dairy production costs, analysts must be careful to account for all relevant costs. Some are explicit and easy to record. For example, farms that purchase feed record feed expenses and quantities. Hired labor is also an explicit cost to operations; the operator incurs a specific expense for the hours worked during any time period.

But significant implicit expenses are also incurred on dairy farms and are much harder to measure. For example, farm operators and their families contribute labor to the dairy enterprise. Although unpaid, the cost of the labor should still be recognized. The operator or family members could have earned income by working off the farm, and their foregone labor earnings represent the opportunity cost of the farm's unpaid labor.

Dairy farms often incur two other important implicit expenses, for homegrown feed and for capital equipment and structures. Homegrown feeds and forage represent implicit costs because the operator could have sold the feeds or the land supporting their production. Many operations own equipment and structures, and do not record an explicit annual expense for their use. But capital use remains an implicit cost to the farm that could have invested the money elsewhere and earned a return on it.

Two other issues pertain in developing cost estimates: joint production and common costs. Dairy production yields a joint product—milk and livestock, the dairy animals that are culled from the herd and sold. If products are truly joint, the costs of producing them cannot be attributed separately to each product, and attempts to do so may simply underestimate the costs of the enterprise. Next, some costs—such as taxes, administrative overhead, and some energy expenses—are borne at the level of the whole farm (they are common to all commodities produced on a farm). Different analytical approaches may have different means of accounting for joint products and common costs, and this may lead to different estimates.

Measuring Dairy Costs with ARMS

USDA's Agricultural Resource Management Survey (ARMS) provides data on input use, expenses, production, and farm characteristics for a large representative sample of U.S. farms (Appendix). The annual survey contains multiple versions, some targeting producers of specific commodities. Two dairy versions underlie our analyses; one collected data from the year 2000 from dairy farms in 22 States, and the other collected data for the year 2005 from dairy farms in 24 States.

The 2005 survey included specific questions targeted at organic dairy operations and a sample design that would ensure adequate statistical coverage of them. About 1 percent of the Nation's dairy cows were certified organic in 2005. Organic operations tend to be smaller than conventional farms, and to have higher expenses and higher revenues per cwt of milk produced. Our 2005 cost analysis excludes organic operations because their cost structure differs significantly from conventional producers. Organic operations were not separately identified in the 2000 data, when they accounted for about 0.4 percent of the nationwide herd, and some organic operations probably appear in that data set.[6]

The ARMS asks dairy producers about cow inventories and milk production, technology choices, structures and equipment, input use and expenses, and manure management strategies and technologies. It also elicits information on revenues, expenses, production, assets, and liabilities at the whole-farm level, as well as information about the farm operator's household.

The survey's information can be combined with additional analyses and data to estimate implicit expenses. ERS staff use off-farm wage data from another version of ARMS to estimate the opportunity costs of unpaid labor hours used on the farm. Market price data, from other USDA sources, are used to value the reported quantities of homegrown feed and forages fed to dairy cows. Finally, ERS analysts produce annualized estimates of the cost of replacing the capital used for cattle housing, milking facilities, feed storage structures, manure handling and storage structures, feed handling equipment, tractors, trucks, and purchased dairy herd replacements, plus the interest that the remaining capital could have earned in an alternative use. ARMS respondents report the type, capacity, and characteristics of different types of equipment and structures in the dairy enterprise. ERS analysts add information on acquisition prices, useful lives of various types of capital, and interest rates to estimate annual capital replacement costs.

What Can ARMS Tell Us about Scale Economies in Dairy Farming?

The ARMS dairy versions provide detailed data for large samples of dairy farms of widely ranging sizes. The data collected offer a powerful resource for analyzing dairy farm costs. Although there have been other studies of dairy production costs, we focus on ARMS-based studies because the data are recent and they encompass large samples across a wide range of farm sizes and locations. Two approaches have been applied to ARMS data: dairy enterprise cost-of-production (COP) accounting and econometric estimates of dairy cost functions.

Cost-of-Production Accounting

COP accounts use detailed data on farm inputs and outputs, drawn from ARMS and external sources, to build estimates of total costs of production and gross returns. ERS develops cost and return estimates for several commodities. This report presents estimates of mean costs and net returns for 4 farm size classes in 2000, based on a sample of 819 farms (table 3), and for 6 farm size classes in 2005, based on a sample of 1,462 farms (table 4). Estimates are expressed in dollars per hundredweight (cwt) of milk produced. More COP documentation is provided on the ERS website, at www.ers.usda.gov/Data/ARMS/CostOverview.htm.

In 2000, mean costs of production fell as enterprise size increased (table 3). For example, average total costs on farms with at least 500 milk cows ($12.39 per cwt) were 18 percent below average total costs on farms with 200–499 cows, a sizeable advantage. Costs were much higher for farms with fewer than 200 milk cows. Thus, there may be important economies of scale in dairy production. Estimated "ownership costs" (particularly for housing, milking facilities, and machinery) fall sharply as farm size increases, suggesting that larger enterprises use their equipment and structures more intensively. Labor costs per cwt of milk also fall quite sharply. Finally, feed costs account for a large share of total costs across farm sizes, but appear not to be a source of substantial scale economies, as average feed costs did not fall sharply with size.[7]

COP estimates for 2005 cover a wider range of size classes, with the largest class in the 2000 data, 500 or more cows, split into two. Average costs of production still fall as herd size increases, and the differences are large. Farms with 1,000 or more cows realized average costs 15.4 percent below those in the next smaller class (500–999 cows) and 24 percent below farms with 200–499 cows. Costs at smaller operations are considerably higher (table 4).

The 2005 data also reveal some sources of cost advantage. Overhead expenses, particularly those associated with capital recovery and with the operators' unpaid labor, still fall sharply as herd size increases. But note that average operating costs also fall noticeably at larger sizes, and the largest farms seem to incur lower total feed costs (purchased plus homegrown plus grazed), per cwt of milk produced, than small operations.[8]

Net Returns

The gross value of production generated by the dairy enterprise includes payments from milk production, from sales of dairy animals, and from other sources (such as leasing of animals or space, dairy co-op patronage dividends, or the value of manure produced). Net returns are the difference between the gross value of production and total costs. Enterprises with positive net returns cover all costs, including costs of capital recovery.

Table 3. Dairy costs of production, by herd size, 2000

	Enterprise size (number of milk cows)			
	<50	50-199	200-499	>499
Mean herd size (*milk cows)*	33	88	313	955
Output per cow *(pounds*)	14,932	16,157	17,420	17,326
	Dollars per hundredweight			
Total operating costs	11.61	9.75	8.49	8.63
All feed	8.16	6.54	5.83	6.17
Total labor costs	11.90	6.04	2.77	1.86
Hired labor	0.32	1.01	1.45	1.41
Unpaid labor	11.58	5.03	1.32	0.45
Total ownership costs	6.88	5.08	3.89	1.90
Housing facilities	1.57	1.31	1.14	0.48
Milking facilities	1.33	0.66	0.10	0.06
Machinery	2.26	1.43	0.54	0.26
Total costs	30.39	20.87	15.15	12.39
Gross value of production	15.74	14.68	14.06	13.41
Net returns	-14.65	-6.19	-1.10	1.02

Source: ERS estimates, at www.ers.usda.gov/data/arms/CostOverview.htm

Herd size refers to all dairy cows on an enterprise, including dry cows but excluding calves, heifers, and bulls. Gross value of production for the dairy enterprise includes milk, cull cattle sales, and other revenue generated by the dairy enterprise. Net returns are the difference between gross value of production and total costs.

Table 4. Dairy costs of production, by herd size, 2005

	Enterprise size (number of milk cows)					
	<50	50-99	100-199	200-499	500-999	>999
Mean herd size	35	69	133	295	666	2083
Output per cow(lbs)	15,055	17,149	18,228	19,487	20,719	20,195
		Dollars per hundredweight				
Total operating costs	12.30	12.94	11.51	11.31	11.07	9.74
Purchased feed	3.60	3.75	4.12	5.00	5.64	5.99
Homegrown feed	5.02	5.07	4.06	3.01	2.58	1.47
Grazed feed	0.41	0.15	0.11	0.10	0.02	0.01
Allocated overhead	17.79	12.56	9.31	6.61	5.00	3.85
Hired labor	0.50	0.80	1.34	1.84	1.80	1.61
Unpaid labor	10.60	6.10	3.13	1.34	0.54	0.17
Capital recovery	5.26	4.56	3.89	2.55	2.03	1.66
Total costs	30.09	25.50	20.82	17.92	16.07	13.59
Gross value of prod.	17.87	17.56	17.20	17.25	16.56	16.54
Net returns	-12.22	-7.94	-3.62	-0.67	0.49	2.95

Source: ERS estimates, at www.ers.usda.gov/data/arms/CostOverview.htm

Herd size refers to all dairy cows on an enterprise, including dry cows but excluding calves, heifers, and bulls. Gross value of production for the dairy enterprise includes milk, cull cattle sales, and other revenue generated by the dairy enterprise. Net returns are the difference between gross value of production and total costs. Organic operations are excluded.

Farms with at least 500 cows had positive net returns, on average, in 2000, while the three smaller classes had negative net returns (table 3). In 2005, farms with 500–999 cows had net returns of 50 cents per cwt of milk, while farms with at least 999 cows had net returns of nearly $3 per cwt; the smaller classes again had negative net returns.

The estimates of net returns are based on national averages, and there is often a wide variation around average performance in agriculture. Some small operations may be exceptionally well-managed, while others may have below-average input prices or above-average product prices. As a result, even though small farms show losses on average, some can be quite profitable. For example, farms with 100-199 head had average net returns of -$3.62/cwt in 2005, but 25 percent of them realized positive net returns, with the gross value of production exceeding total costs, including the opportunity costs of capital and operators' labor (Figure 3). Six percent of farms with 50–99 head earned positive net returns, as did 41 percent of farms in the 200–499 class.

Net returns drive entry and expansion decisions: farmers are unlikely to commit capital and labor to new projects that are unlikely to cover the costs of those decisions. But farms that are already in business and are considering whether to continue operating make another judgement. Those operators have already committed their equipment and structures, and that sunk capital may be an opportunity cost that is virtually zero—the salvage value. Capital recovery costs may, therefore, be irrelevant to the decision to continue operating. What matters in that case is not whether gross returns exceed total costs, but whether gross returns exceed the farm's operating costs, plus the opportunity costs of the operator household's labor.

To get at that measure, the share of farms whose gross returns exceed all costs except for capital recovery must be calculated. Fifty percent of farms with 100–199 cows meet that standard for profits, as do 25 percent of those with 50–99 cows and 73 percent of those with 200–499 cows (Figure 3). These farms will likely continue to operate because they cover their immediate costs, including the opportunity costs of operator labor.[9]

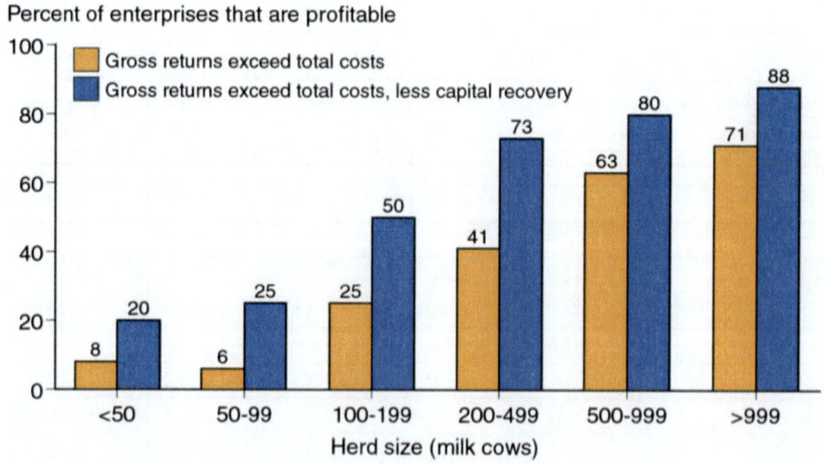

Source: ERS estimates, from 2005 ARMS dairy version.

Figure 3. Profitable dairy enterprises were more common among large farms in 2005

A substantial share of smaller dairy farms seems to earn enough from operations to keep operating, and in some cases to be quite profitable. But, on average, farms in smaller size classes are not covering the opportunity costs of their investment in capital and the operator's time. Correspondingly, large dairy farms are returning profits in excess of the owners' time and capital costs. The differences in estimated returns mirror the changes in structure—production is shifting away from smaller farms, toward much larger dairy farms. Because many existing smaller operations are economically viable and will remain so for a long time, structural changes will play out over an extended period of time.

Looking to the Future: Expected Structural Changes

The data show wide disparities in net returns across farm size classes, suggesting that structural shifts toward large operations will likely continue. We can use ARMS data to develop a forward-looking analysis of survival expectations among existing farms, as a check on the net returns findings. The 2000 and 2005 ARMS dairy versions asked respondents for the number of years that they expected their present operation to continue producing milk.[10]

Exit expectations have a strong inverse association with herd size. In the 2000 survey, over 30 percent of operators with under 100 cows expected their operation to end milk production by 2005, and over 50 percent by 2010.[11] By contrast, less than 4 percent of operators in the largest size class (500 or more cows) expected their operations to end milk production by 2005, and about 15 percent by 2010 (table 5).

Table 5. Prospective exit by dairy farms

Herd size		Percent of operations ending production:	
	Sample observations	Within 5 years	Within 10 years
2000 ARMS			
1-49	54	39.0	59.0
50-99	416	30.7	57.0
100-199	186	21.4	47.0
200-499	87	13.9	35.2
>499	76	3.7	15.5
2005 ARMS			
1-49	164	35.5	69.5
50-99	289	26.1	48.2
100-199	347	18.5	43.1
200-499	336	10.3	29.3
500-999	179	8.2	20.7
>999	147	7.4	22.0

Source: Agricultural Resource Management Survey (ARMS), 2000 and 2005 dairy versions. Tabulation of responses to the question "How many more years do you expect this operation to continue producing milk?" Respondents chose among less than 1 year, 1 year, 2-5 years, 6-10 years, 11-19 years, and 20 or more years. Organic operations are excluded.

The 2005 responses show the same strong inverse relationship. Nearly 36 percent of operations in the smallest size class, and over 25 percent in the 50–99 head class, expected to end milk production by 2010, with much higher shares expected to leave by 2015. In contrast, only 7 percent of the largest operations expected to close by 2010. The largest class is open-ended (1,000 or more head), and exit expectations are inversely related to size within the class as well, with exit expectations concentrated among operations with less than 2,000 head (28 percent of operations with 1,000–1,999 head expect to close by 2015, compared to 10 percent of those with 2,000 or more). The survey responses support the implications for structural change based on costs and returns—while many small dairy farms are economically viable and will remain in business, many others will exit, and production will continue to shift to large farms.

Milk prices can fluctuate sharply. The average farm-level milk price in 2005 was $15.14 per hundredweight. Prices fell to $12.90 in 2006 before rising to hit $20 in June 2007. In turn, this year's sharp price increases were driven by increased world demand for dry dairy products, lowered production subsidies in some countries, and ethanol-fueled increases in feed prices. At the prices realized in summer 2007, more small and midsized dairies will be financially viable. But milk demand and milk prices will continue to fluctuate, and the cost relationships outlined here have not been fundamentally altered. Larger operations still have substantial cost advantages, and shifts of production to larger operations will place downward pressure on industrywide costs and prices, thus offsetting some of the impact of any long-term increases in feed expenses.

Behind Net Returns: Revenues and Farm Size

The prices that farmers receive for their milk vary by region and are higher in those regions with more small operations. For example, milk prices ranged from around $14.00 per cwt in California and Idaho, where production is concentrated in large farms, to $15.60 in Wisconsin, $15.90 in New York, and $16.90 in Pennsylvania, where small farms still predominate (using 2005 USDA/NASS data on average annual prices received, for all milk). Revenues from milk sales account for most of the gross value of dairy production—89 percent, on average, across all 2005 sample farms—with revenues from the sale of dairy animals accounting for most of the remainder. Hence, the gross value of dairy production varies systematically across size classes, with smaller operations holding an advantage.

While milk prices tend to be higher in regions with smaller dairies, the variation in prices received across regions is far lower than the variation in costs across farms of different sizes. As a result, the price advantage that small farms gain by operating in Pennsylvania or Wisconsin will rarely be enough to offset their cost disadvantages. In addition, large dairy farms that locate in regions with higher prices can gain that revenue advantage and still realize scale economies. While the pattern of prices may have allowed some small and midsize producers to stay in operation longer, size-related price and revenue differences are much smaller than the cost differences that appear to be driving structural change.

Gross value of production does not include government payments. The 2002 farm bill introduced countercyclical payments under the Milk Income Loss Contract (MILC) program; farmers could receive direct payments in months when market prices fell below a target level.

Specifically, the payments were equal to 45 percent of the difference between $16.45 and the reference market price for milk (the Federal Milk Marketing Order Class 1 price at Boston), when the reference price was lower. The program expired in September 2005, and was extended by Congress for 2 years, with payments reduced to 34 percent of the difference between the target and reference prices.

Payments are restricted to the first 2.4 million pounds of production on a farm. While farms of all sizes are eligible for payments, total annual payments are capped at the amount that would be provided to a producer with a herd of about 120 cows (at 2006 average milk yields). MILC payments therefore provide substantially more support, as a proportion of gross receipts, to smaller operations.[12] Payment rates on eligible production have ranged as high as $1.82 per cwt (in April 2003), but remained at zero for most of 2005, as the reference price remained above the target. While government payments under the MILC program are concentrated on small dairy operations, net returns at most small dairies remain negative even after accounting for MILC payments.

Because many small dairy farms operate near the margin of viability, enhanced revenues—from higher product prices, countercyclical support, or value-added activities such as agri-tourism or cheese-making—may sustain these operations. Other small operations may be able to adopt production technologies, such as managed grazing, that lead to lower gross returns, but substantially lower costs. Still others have turned to organic production, which offers higher milk prices (along with higher feed costs). Regardless, continued shifts of production to larger enterprises will place downward pressure on conventional milk production costs and prices, and that will impose powerful competitive pressures on small operations and on alternative products and production technologies.

Behind Net Returns: Estimates of Unpaid Labor Expenses

One major component of the small farm cost disadvantage is the opportunity cost of unpaid labor provided by operators and their families (table 4), which forms a much higher share of total costs in small farms than in large. Because our estimates of unpaid labor expenses loom large in cost differences, we examine the estimates more closely.

Unpaid labor expenses reflect the amount of unpaid labor provided, and the implicit opportunity cost of that labor. The ARMS questionnaire obtains information on unpaid hours provided by the principal operator, other operators, family members, and others. Total unpaid hours do not vary greatly with herd size and, as a result, unpaid hours per cwt of milk produced fall sharply as the volume of milk produced on the farm increases (table 6).

The opportunity cost of unpaid labor is based on the off-farm labor earnings of all farm households (El-Osta and Ahearn, 1996). Version 1 of the ARMS ascertains annual hours worked off-farm, and off-farm wages and salaries earned, by responding principal operators and their spouses. ERS analysts then use statistical regression analyses to identify how hourly off-farm earnings for all farm operator households vary, for operators and for spouses, by age, education, and location.[13] The results can then be used to estimate the off-farm wages that dairy operators and spouses—with specified age, education, and location characteristics—gave up by working on the dairy enterprise instead of off the farm.[14]

Table 6. Drivers of unpaid labor expenses, 2005

	Enterprise size (number of milk cows)					
	<50	**50-99**	**100-199**	**200-499**	**500-999**	**>999**
	Annual means, by size class					
Production (*cwt*)	5,213	11,828	24,218	57,539	138,071	420,665
Unpaid hours						
Principal operator	2,376	3,095	3,124	3,111	3,150	2,987
All	3,339	4,190	4,372	4,111	3,742	3,450
Hours/cwt	0.64	0.35	0.18	0.07	0.03	0.008
Mean hourly wage (*$*)	16.85	17.50	17.58	18.89	19.53	20.55

Sources: Production and hours, as reported in Agricultural Resource Management Survey, 2005, version 4 (dairy). Mean hourly wage, imputed by ERS on the basis of statistical analysis of off-farm earnings reported in 2005 ARMS, version 1. Organic operations are excluded.

Estimated opportunity costs of off-farm labor varied across dairy farms in the 2005 survey, depending on location, the shares of total unpaid hours provided by operators and by spouses, and their ages and education. The mean off-farm hourly wage applied to dairy enterprises was estimated to be $17.58 per hour, and it varied from $15.08 at the 10th percentile (10 percent of farms had lower values) to $20.74 at the 90th percentile. It also varied systematically across farm sizes, from a mean of $16.85 among farms in the smallest size class to $20.55 in the largest class (table 6). Even though the estimated opportunity cost of unpaid labor was higher in the larger farms, the cost of unpaid labor per cwt of milk produced was much higher at smaller farms because they use much more unpaid labor per cwt.

Small farm production costs look more competitive with large farms if the opportunity cost of unpaid labor is ignored, if the earnings that unpaid labor could have obtained off the farm are lower, or if lower unpaid hours are reported. In turn, some small dairy operators may continue to operate, even at an estimated loss, because they are willing to accept less than they can earn in nonfarm employment. However, even with substantial changes in the estimated opportunity costs of off-farm labor, small dairy farms' costs would still, on average, be well above large farm costs and fall well below small farm revenues. For example, suppose the average opportunity cost of labor were to fall to $10 an hour, well below the estimate of $17.50 for farms in the 50–99 size class. Unpaid labor expenses would fall to $3.49 per hundredweight, from $6.11, and total costs would fall to $22.97, still above the average gross value of production of $17.80 for farms in this size class, and far above large farm costs of $13.60.

How Does Structural Change Affect Industrywide Costs?

Shifts to larger enterprises, by allowing for scale economies, lower average dairy production costs. We can estimate the impact of recent structural change on industrywide costs by averaging the cost estimates for enterprises in different size classes (table 4), using weights drawn from the 2000 and 2006 size distribution of U.S. milk production (shown in table 1). We first calculate a weighted-average COP, using table 4 total cost estimates and weights reflecting the distribution of production in 2000, of $ 19.83/cwt. If we then recalculate using weights reflecting the 2006 distribution of production, the weighted-average COP falls 8 percent to $18.24. Many factors affect actual costs of production, but this is a sizeable impact in a short span of time. The longer run impacts of structural change could be quite substantial.

WHAT DO ECONOMETRIC ESTIMATES TELL US?

COP accounting estimates clearly show that average costs decline as herd sizes increase, and they provide some useful information for assessing the sources of the cost advantage, but they also have limitations. Specifically:

- Because the estimates do not distinguish between input quantity and input price, we cannot determine whether a cost advantage derives from more efficient input use or from lower prices paid.
- COP estimates reflect the average performance of farms in each size class. Farms vary in efficiency—some are best-practice efficient operations, while others may be poor performers. Consequently, costs can fall as herd sizes increase, either because larger enterprises tend to be more efficient or because technology creates scale economies that allow large enterprises to realize lower costs than equally efficient smaller enterprises.

Two econometric analyses estimate scale economies in dairy production with data from the 2000 ARMS dairy version (Tauer and Mishra, 2006; and Mosheim and Lovell, 2006). The studies take different approaches to the issue (see box, "Herd Size and Production Costs: Scale Economies or Inefficiency?"). Each finds that average production costs fall as herd sizes increase, and each aims to identify the roles of scale economies and inefficiency.

Tauer and Mishra (T&M) argue that most of the observed cost advantage of large herds follows from a greater incidence of inefficient production among smaller dairies. Scale economies were found to be quite modest once they accounted for inefficiency. Costs at fully efficient large dairies (1,000 milk cows) were estimated to be only $1.13 per cwt, or 11.3 percent, below those at fully efficient small (50 cow) dairies, in contrast to an $8.10 difference (36.8 percent) using unadjusted 2000 data. They estimated that average costs at efficient dairies with 1,000 cows were 3 percent below those of efficient dairies with 500-cow herds, versus a 14.3-percent difference based on unadjusted data.

Mosheim and Lovell (M&L) found scale economies to be much more important. In M&L's analysis, average costs among efficient producers decline sharply as herd size expands to 400 milk cows, and they continue to decline, but less rapidly, beyond that size (Figure 4). Among the most efficient operations, average costs fall to $10.57 per cwt at 2,400 head, compared with estimates of $11.05 at 1,300 head, $12.43 at 700 head, and $18.25 at 300 head. Furthermore, while the estimated cost advantages of further increases in herd size are modest at sizes above 1,000 head, M&L find that scale economies are not completely exhausted even among the largest operations in the sample (2,000-3,000 head).

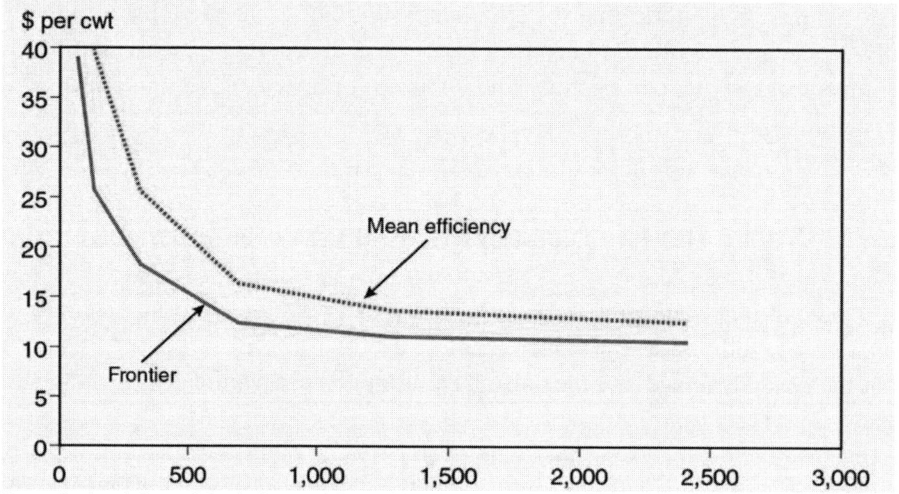

Source: Data derived from Mosheim and Lovell (2006).

Figure 4. Estimated scale economies in dairy production

M & L also found that inefficiency was an important source of cost differences. As in T&M, inefficiency was more prevalent among smaller operations. Costs for the average very large farm (2,400 head), at $12.55 per cwt, were 19 percent above frontier costs (Figure 4). Costs at average (mean efficiency) farms are 32 percent above frontier costs at 700 head, and 40 percent greater at 300 head.

HERD SIZE AND PRODUCTION COSTS: SCALE ECONOMIES OR INEFFICIENCY?

The cost curve in figure 2 (p. 6) reflects how costs vary among producers who are choosing and using inputs in such a way as to minimize costs. Such producers are allocatively efficient in that they are choosing the combinations of inputs that will allow them to minimize the costs of producing a given level of output, and they are productively efficient in that they are getting the most out of the inputs that they've chosen. In that case, the declining cost curve represents scale economies that allow costs for efficient producers to decline as output expands. Scale economies are a technological concept, and in dairy production they may arise from several sources, including milking systems and milk storage, housing, feed storage and delivery systems, and manure handling equipment.

Inefficient operations would fall above the cost curve in figure 2, either because they are allocatively or productively inefficient. Operations can be inefficient because of events outside of the operator's control, such as bad weather; because the operation was originally designed and built to take advantage of input prices that no longer hold; or because of poor decisions made by the operator. The wide range of costs and returns exhibited by dairy farms in the COP estimates, as in other analyses of farm performance, strongly suggests that there may be important differences in efficiency among farms.

Analysts seeking to distinguish scale economies from inefficiency aim to identify the cost line depicted in figure 2. In principle, inefficient enterprises would have costs above the unit cost line, while efficient dairy enterprises would be on the line. Actual data points can fall above or below the line for other reasons, such as measurement errors in the data or an inability to control for other factors that affect costs. These are called random, or stochastic, errors. In trying to identify the unit cost line (scale economies) in the data, and to identify the extent of inefficient production, assumptions are made about the nature of the stochastic errors and about the nature of the technology that drives the shape of the line. The two analyses of the 2000 ARMS dairy data took different approaches to modeling and data development, and these differences affect their conclusions.

Tauer and Mishra (T&M) imposed two assumptions that are likely to reduce the estimate of scale economies in their analysis. First, they subtracted culled cattle revenues from the ERS cost-of-production estimates, on the grounds that those revenues represent the separable costs of livestock production and that they wanted to focus on the specific costs of milk production alone. But milk and cull cows are joint products, so costs cannot be meaningfully separated. Moreover, there is a strong inverse relationship between culled cattle revenues, per cwt of milk sold, and herd size in the sample. Thus, deletion of livestock sales from costs will reduce estimated production costs, and will reduce them more among smaller operations.

Second, T&M did not control for input prices in their analysis, but instead controlled for the locations (States) of farms (by inserting a dummy variable for each State). The practical impact of that approach is to limit the effective range of farm sizes considered to the range within a State—cross-State differences in farm size, which are large, were not used to assess scale effects. The adjustment for culled cow revenues and the imposition of State effects are each likely to reduce estimated scale economies.

Mosheim and Lovell (M&L) took a different conceptual approach to modeling costs. Costs at the level of the whole farm were analyzed, and the impact on costs of changes in all farm production (milk, but also crops and other livestock), as well as in input prices, was investigated. This approach is theoretically more appropriate than COP accounting, since it does not rely on potentially arbitrary rules for assigning joint or common costs to different farm enterprises, but it also presents significant technical and reporting challenges.

M&L developed an extensive set of input prices, and included those prices in their analytical model. The model allowed for a flexible specification of the relation between the scale of output and costs, and also accounted for inefficiency among producers. Cost data were drawn from ERS ARMS files, but two important expense categories were adjusted.

M&L used a different approach to estimating the implicit cost of capital equipment and structures used on the farm. The COP analyses build an estimate of capital stock by using detailed ARMS survey data on the structures and equipment used in the dairy enterprise, and estimating capital recovery costs from that information. M&L estimate costs for the whole farm, not just the dairy enterprise, in order to better model the impact of joint and common costs. The survey does not contain structures and equipment detail for the whole farm, so M&L estimate the farm's capital stock using data on estimated capital prices and a farm's financial flows.

M&L's estimates of the opportunity cost of unpaid farm labor exceed COP estimates. Since unpaid labor is more important on smaller operations, this approach raises estimated costs on small farms compared with COP and T&M's estimates, and raises the estimates of scale economies. The COP estimates are based only on off-farm wage earnings. M&L take the same regression-based approach that is used for the COP estimates, but they add earnings from operating another business to off-farm wage earnings. Since M&L include more sources of income in their analyses, their estimates of the costs of unpaid labor are higher.

USING INVESTMENT DATA TO SUPPLEMENT COST ESTIMATES

In the ARMS data, average costs of dairy production fall sharply as herd sizes increase. Econometric analyses of the data strongly suggest that there are scale economies in dairy production, and there is evidence that costs continue to fall as herd sizes pass 1,000 head. But the econometric studies differ in their estimates of the magnitude of scale economies, and hence of the driving forces behind the cost-size relationship. T&M attribute most of the large cost difference between large and small dairies to widespread inefficiency among small producers. M&L find more inefficiency among small producers, but also find that scale economies create substantial cost advantages for large dairies.

In addition, the 2000 ARMS dairy sample contained few farms with more than 2,000 cows. While it is much more extensive, the 2005 sample is also thin for very large sizes. As a result, the data cannot be used to provide reliable estimates of cost differences among very large dairy farms, and we cannot learn much about maximum efficient scale.

Because the ARMS data provide only limited observations for tracking costs among large operations, and because the econometric estimates offer conflicting estimates of the importance of scale economies, we supplemented the analyses by tracking capacity additions.

We use an approach known as a "survivor analysis" (Stigler, 1958), which is based on the assumption that owners will build new plants at sizes that realize the lowest costs for the era in which they are built (that is, for the existing technology and input prices). If that is true, then one ought be able to infer the efficient range of plant sizes by analyzing the investment decisions made by operators of new plants. The approach does not require detailed cost information, but does require complete information on the size distribution of herds in each of several periods.

Fortunately, we have such information available from census of agriculture records (Appendix). We use data from the 1992, 1997, and 2002 censuses to analyze changes in the distribution of dairy farm sizes.[15] The total number of dairy cows fell from 9.5 million in 1992 to 9.1 million in 1997, where it remained in 2002 (table 7). But cows shifted to much larger farms. Farms with fewer than 200 milk cows accounted for 68 percent of all dairy cows in 1992, but lost nearly 2.5 million milk cows over the next 10 years as their share of total cow inventory fell to 44 percent. Meanwhile, farms with at least 1,000 cows gained over 1.7 million cows as their share of national inventory grew from 10 to 29 percent. Those investment decisions are entirely consistent with the COP findings (tables 3 and 4) that larger farms generally achieved economic profits (returns above the costs of capital), while farms with fewer than 500 cows incurred economic losses.

To learn more about behavior among large dairies, all farms with at least 1,000 milk cows were sorted into more detailed size classes, and size distributions for each census year were compared. Dairy farms with at least 1,000 cows added 653,000 dairy cows during 1992–

97. Farms with 1,000–1,999 head added 43 percent of that total, and farms with 2,000–2,999 head added another 28 percent (Figure 5). The remaining 29 percent of the added inventory, or 190,000 cows, was added by farms with at least 3,000 head. Large dairy farms in 1992–97 generally had between 1,000 and 3,000 cows.

The pattern changed sharply in 1997–2002. Large dairies as a group added over 1 million cows to their inventory, but nearly 70 percent (or 723,000 head) now went to farms with at least 3,000 head. Farms with at least 5,000 head added over 250,000 cows, four times more than the number added in 1992–97.

Investment in large new dairies is occurring in all major production regions, although the very largest still tend to be in the West (table 8). In 1992, there were only 15 dairy farms with at least 1,000 milk cows in traditional dairy States, compared with 483 in Western States. There were 976 large Western dairy farms by 2002, with farms of 3,000–10,000 head growing more prevalent. But the number of large dairy farms also expanded rapidly in traditional States as well, to 67 in 1997 and to 178 by 2002.

Table 7. Changes in cow inventory, by farm size, 1992-2002

Herd size	Number of milk cows (1,000)			Change, 1992-2002 (1,000)	
	1992	1997	2002	1992-1 997	1997-2002
0-199	6,497	5,186	4,028	-1,311	-1,168
200-499	1,302	1,395	1,336	+93	-59
500-999	756	924	1,115	+168	+191
>999	937	1,590	2,651	+653	+1,071
All farms	9,492	9,095	9,130	-397	+35

Source: Census of agriculture, for reported years. Herd size refers to all dairy cows on an enterprise, including dry cows but excluding calves, heifers, and bulls.

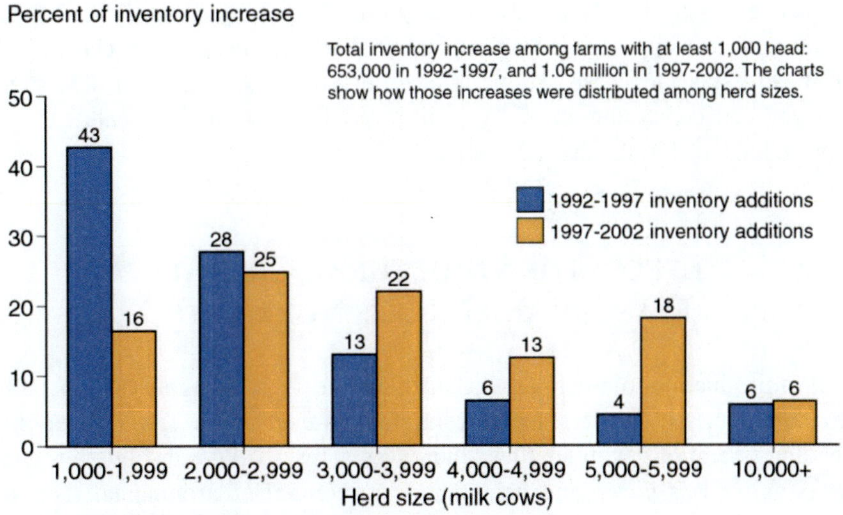

Source: Census of agriculture farm records.

Figure 5. Inventory additions shifted to farms with over 3,000 head after 1997.

Table 8. Large dairy farms, by region and size class

Region and herd size	1992	1997	2002
		Number of dairy farms	
Western dairy States			
1,000-1 ,999 cows	401	545	606
2,000-2,999 cows	60	132	197
3,000-3,999 cows	9	40	94
4,000-4,999 cows	8	16	40
5,000 or more cows	5	12	39
Traditional dairy States			
1,000-1 ,999 cows	12	59	135
2,000-2,999 cows	3	7	27
3,000-3,999 cows	0	1	12
4,000 or more cows	0	0	4
Other States			
1,000-1,999 cows	37	50	54
2,000-2,999 cows	10	7	25
3,000-3,999 cows	5	8	9
4,000-4,999 cows	1	3	4
5,000 or more cows	3	3	8

Source: ERS tabulations, from census of agriculture microdata.

Western dairy States: AZ, CA, CO, ID, MT, NV, NM, OR, TX, UT, WA, and WY.

Traditional dairy States: CT, DE, IA, IL, IN, MA, MD, ME, MI, MN, MO, NH, NJ, NY, PA, OH, RI, VT, and WI.

Other States: AK, AL, AR, GA, FL, HI, KS, KY, LA, MS, NC, ND, NE, OK, SC, SD, TN, VA, and WV.

The data show sharp shifts of new capacity to very large dairies, with 3,000 to 10,000 head, and a slackening of investment in dairies with 1,000–3,000 head.[16] The capacity shifts mirror the data on exit expectations among large dairies, where 28 percent of those with 1,000-2,000 head expect to cease production within 10 years. This shift also suggests that there may have been constraints on dairy farm size (diseconomies of scale), but that those constraints have loosened in the last 10 years.[17]

Effects of Structural Change: Manure and Excess Nutrients

The shift of production to much larger dairy farms, driven by significant cost advantages, is likely to continue. The shift also concentrates production in fewer locations and on confined feeding operations with smaller land bases. This spatial consolidation concentrates animal wastes as well as cows, and heightens the potential environmental risks associated with milk production. Here, we describe the pace of geographic consolidation in dairy production and evaluate the potential impact on excess nutrients applied to the land. We then review the environmental regulations developed to deal with dairy farms and other confined animal feeding operations, and summarize ARMS-based data on how dairy farms are managing their manure.

Geographic Consolidation of U.S. Dairy Production

What do we mean when we say that dairy production is spatially concentrating? First, although more milk is being produced then ever before, the industry is consolidating into fewer dairy counties. We sorted all U.S. counties according to the number of dairy cows in each, using census of agriculture data. We then determined the number of counties necessary to account for one-quarter of all dairy cows, and the number necessary for half.

The changes over time are striking. In 1969, 71 counties accounted for one- quarter of all dairy cows, but that count fell to 34 counties in 1992 and just 20 by 2002. Correspondingly, 247 counties accounted for half of all dairy cows in 1969, versus 130 in 1992 and 95 in 2002.

Larger dairy farms concentrate their herds on a more limited land base. Figure 6 shows how cow density varies with herd size for farms in traditional dairy States, where crop production and pasture have historically been combined with milk production. The figure shows two measures of density: cows per 100 acres of farmland and cows per 100 acres of cropland. For each measure, density rises as herd size increases, and densities in the largest class are twice those in the next largest.[18]

As dairy production consolidates geographically, the associated consolidation of manure may lead to the production of manure-based nutrients, primarily nitrogen (N) and phosphorus (P), in excess of what can be taken up by crop production on the dairy farm's land. If improperly managed, excess N and P applications can pose environmental and human health risks. That's not a necessary outcome of geographic consolidation: the farm might be able to remove manure for spreading on other operations or treat the manure to limit environmental harm. Before we consider manure management practices, we will next evaluate the potential changes in excess manure-based nutrients associated with dairy farm consolidation.

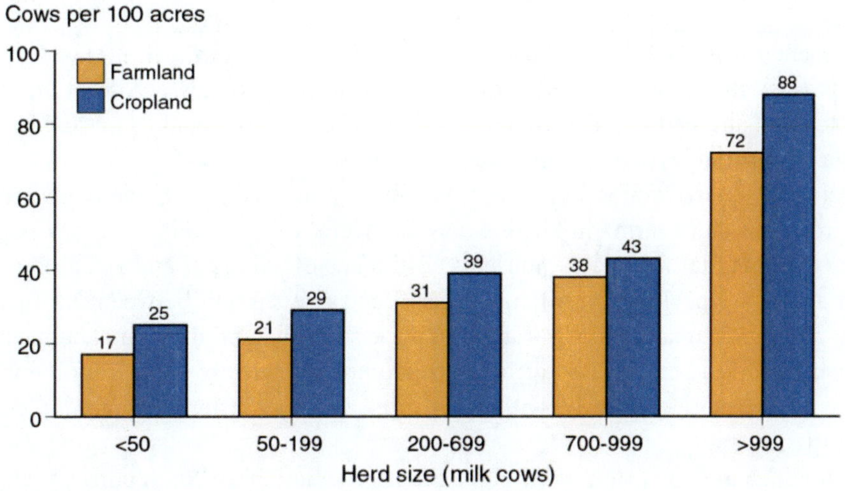

Source: 2005 ARMS dairy version.

Figure 6. Cow density increases with herd size

FARM SIZE AND THE POTENTIAL FOR EXCESS NUTRIENTS

The N and P available for crop production on a farm consists of the quantities of those nutrients found in the manure generated on the farm, along with the level of commercial fertilizers placed in the ground. Only certain levels of these nutrients are required to grow crops. Excess nutrients, if not properly managed, can build up in the soil and contribute to groundwater contamination, excess runoff to streams and rivers, and air pollution. We estimate the nutrients present in livestock manure on dairy farms, and compare them to the nutrients required by the farms' mix of crops, to show that large dairy farms can produce substantial excess nutrients from manure alone.

The analysis is limited to farms in traditional dairy production regions—the Northeast (New York, Pennsylvania, and Vermont) and the Midwest (Iowa, Illinois, Indiana, Michigan, Minnesota, Missouri, Ohio, and Wisconsin). Dairy farms in these regions typically raise crops and spread manure on farmland. With the region's production moving toward large industrialized operations, there may be increased stress on limited land bases.[19]

In developing measures of excess manure-based N and P, we follow the conceptual approach used in earlier ERS research (Gollehon et al., 2001; Ribaudo et al., 2003). We began with year-end livestock inventories for each of the four major species in the States we examine—beef cattle, dairy cattle, hogs, and poultry. To convert onfarm livestock inventories to estimates of manure production, we then applied the Kellogg et al. (2000) estimates of annual manure production, in tons, for animals of each species.[20] The nutrient composition of manure varies across species, and the same source provides species-specific conversion factors to construct estimates of N and P from each ton of manure for each species.

Next, the ARMS database describes each sample farm's crop mix—the acreage devoted to each crop that the farm produced and production, in bushels, from that acreage. We used that data to develop farm-level measures of total N and P use by crops, relying on the estimates of assimilative capacity, in pounds of N and P per bushel, for each crop provided in Kellogg et al. (2000).[21] Finally, we subtracted onfarm nutrient use from manure-based nutrient production to generate estimates of potential manure-based excess nutrients on each farm.

We present mean values of those estimates, sorted by farm size class, in table 9. Superscripts refer to statistical tests of difference in excess pounds across herd sizes; each letter denotes that the entry is statistically different from the estimate reported in the cited column, at a 90-percent level of confidence.

On average, manure production generates nutrients that exceed the amounts required by the crops grown on dairy farms in all size classes. Moreover, consolidation into larger farms exacerbates the potential for excess nutrients. Estimates of excess N and P, based on manure production alone, are positive for all size classes, and rise sharply between the smallest and largest size classes. A great deal of variation in excess nitrogen production occurs within each size class, so that most across-class differences are not statistically significant. However, the phosphorus estimates rise sharply with herd size, and the differences across classes are statistically significant.

These findings are consistent with those reported in another ERS report, which used the 2000 ARMS dairy version to develop measures of potential excess nutrient production from manure (Ribaudo et al., 2003). That study took a different approach to measurement, estimating the amount of land needed for manure spreading in order to meet a goal of zero excess P (and alternatively, zero excess N). It then compared the needed acreage to the acreage over which the farm was actually spreading manure, and to the total acreage on the farm (since farms don't always spread manure on all their land). Most small dairy farms in traditional States had land available to meet nutrient standards, but few farms in the largest size class did.

Most farms do not rely on manure alone for crop nutrients; manure is costly to transport to fields, and often does not contain the appropriate combination of nutrients for specific crops and fields. Since farms also apply commercial fertilizers to crops, our estimates (table 9) understate the potential excess nutrients from manure and commercial fertilizer applications. While dairy farms of all sizes have the potential for substantial excess nutrient production, the potential appears to increase noticeably among larger dairy operations, particularly for phosphorus and as herd sizes exceed 1,000 cattle of all types. As dairy farming continues to consolidate into larger operations, this problem will likely become more widespread.

Table 9. Herd size and excess manure-based nutrients

Nutrient	Herd size (number of cows)				
	<50 (a)	**50-199 (b)**	**200-699 (c)**	**700-999 (d)**	**>999 (e)**
	Excess nutrients (pounds per cropland acre)				
Nitrogen	20[bce]	30[a]	24[ae]	39	54[ac]
Phosphorus	2bcde	7acde	1 0abde	1 6abce	29abcd

Notes: The superscripts refer to the results of statistical tests of difference between columns. All tests are expressed at a 90-percent level of confidence. A lettered superscript denotes that the value reported in a column is significantly different from that in the superscript column.

Source: Agricultural Resource Management Survey (ARMS), version 1, 1996-2002, for 5,183 farms with dairy cows in IA, IL, IN, MI, MN, MO, NY, OH, PA, VT, and WI.

Regulations

Dairy farms, as well as other livestock and poultry operations, are subject to regulation of their manure management practices by Federal, State, and local governments. Regulatory efforts often focus on large operations.

Under the Clean Water Act of 1972, the Environmental Protection Agency (EPA) issued regulations in 1974 and 1976 that established effluent limitation guidelines for large feedlots. In 1999, EPA and USDA published a rule proposing a unified national strategy to limit the environmental impacts of animal feeding operations (AFOs). Following widespread discussion, proposals, comments, and analyses, a revised rule was published in February 2003, effective in April 2003, consisting of a set of regulatory requirements aimed at concentrated animal feeding operations (CAFOs). However, following a legal challenge to the 2003 rule, EPA was remanded to revise some portions of the regulations.[22]

An AFO is defined as an operation that confines animals for at least 45 days in a 12-month period with no vegetation in the confinement area. A CAFO is simply a concentrated

AFO. The operation's size, location, means of wastewater conveyance, site characteristics, and other risk-related issues all factor into the authority's decision to categorize an operation as an AFO or a CAFO.

There are three main classes of CAFOs—large, medium, and small. All dairy operations with at least 700 cows are considered large CAFOs. A medium dairy CAFO has 200–699 cows and either discharges animal wastes directly, or has some manmade device (e.g., ditch, flushing system, etc.) that allows animal wastes to discharge directly into U.S. waters. A small dairy CAFO has fewer than 200 cows but is found by a local permitting authority to be a significant contributor of pollutants in the area.

The final ruling established guidelines that CAFOs must adhere to. Under the original 2003 regulations, all CAFOs had to apply for National Pollutant Discharge Elimination System (NPDES) permits, in effect recognizing all CAFOs as point sources of pollution.The revised rule requires only some CAFOs to obtain an NPDES permit—those that either discharge or propose to discharge animal wastes into U.S. waters. Unpermitted discharges are not allowed, except for agricultural stormwater. All CAFOs, permit holders as well as those obtaining the stormwater exemption, had to have a nutrient management plan (NMP) in place by July 2007.

To the extent possible, NMPs must meet nine minimum elements.[23] They must document:

(1) Adequate onsite waste storage;
(2) Proper management of all animal mortalities;
(3) Diversion of clean water, as appropriate, from any production areas on the operation;
(4) No direct contact between confined animals and any waters of the United States;
(5) Proper disposition of all chemicals and other contaminants used onsite;
(6) Site-specific conservation practices to be implemented;
(7) Protocols for the appropriate testing of the waste generated on the farm;
(8) Protocols to land-apply waste in an acceptable fashion; and
(9) Proper documentation of implementation and management of all the elements.

While these nine elements comprise the minimum necessary to operate in the new regulatory environment, more detailed requirements exist for CAFOs depending on their size, with larger CAFOs subject to more stringent guidelines than smaller ones.

EPA rules set minimum standards that CAFOs must meet. States implement the EPA regulations and may also set higher standards or pursue additional strategies. States have imposed minimum facility setback requirements to separate facilities from their neighbors and limit their odors. Some have also imposed tighter land application rules for manure that may apply to broader classes of farms.

Manure Management Strategies and Structural Change

Farms that cannot meet nutrient standards with their current land base have several options. They could reduce the size of their dairy operation. If this were the least costly option, it would lead to a change in farm structure away from large dairy farms. However, given the substantial production cost advantages to size, this would be a likely choice only if the alternatives were quite expensive.

Farmers could also expand the land base for spreading manure on crops, either by acquiring more land and expanding the farm's cropping enterprise, or by selling the manure to others, giving it away, or paying to have it removed.

Manure contains a lot of water, making it very costly to transport, but farmers can reduce those costs with treatment strategies that aim to separate manure solids from liquids. Solids can be more easily transported and applied to cropland or converted to garden fertilizers and other processed products, whereas liquids remain to be applied to onsite cropland. Treatment may also separate methane gas, used for power generation, from manure. Farmers can also reduce the amount of nutrients in manure, and the amount of manure produced per cow, by altering the feed provided to cows.

Ribaudo et al. (2003) analyzed the costs of removal, which may be the most likely mitigation strategy. They evaluated the increase in production costs that would arise from meeting different nutrient standards (N-based and P-based) for farms in three broad size classes and two broad production regions, under different assumptions of the willingness of other farms to accept manure. In general, the estimated effects on production costs were small. If 20 percent of nearby crop producers were willing to accept manure, production cost increases would range from 0.5 to 3.5 percent. Furthermore, while larger farms would see greater percentage increases in production costs, the differences were modest. The smallest farms (less than 425 head, in this analysis) in traditional dairy States would see cost increases of around 0.5 percent, while farms in the largest size class (more than 1,425 head) would see increases of 1.5 (N-based standard) to 3.5 (P-based standard) percent.[24] Given the production cost differences among farms in different size classes (table 4), the analysis suggests that the cost impacts of meeting N- or P-based standards are unlikely to alter the path of structural change.

We used data from the 2005 ARMS dairy version to ascertain the manure management strategies that dairies are following, particularly movement offsite (table 10). For ease of presentation and to generate useful sample sizes for some questions, we classified farms broadly by herd size (1–699 cows versus 700 or more, the cutoff used to define large CAFOs) and by region (Western dairy States and traditional dairy States).

Table 10. Manure management practices, by region, 2005

	Western dairy States		Traditional dairy States	
	0-699 cows	700 or more cows	0-699 cows	700 or more cows
	Percent			
Manure removal				
Farms removing manure (%)	22	57	5	27
Of the farms that remove:				
Percent of manure removed	17	47	2	15
Percent removed that is:				
Sold	16	17	42	34
Given away for free	49	48	53	41
Hauled away for a fee	35	35	5	25
	100	100	100	100

Table 10. (Continued)

	Western dairy States		Traditional dairy States	
	0-699 cows	700 or more cows	0-699 cows	700 or more cows
	Percent of farms			
Other practices and assets				
Have a CNMP[1]	70	45	26	91
Use a manure separator	6	16	2	12
Use an anaerobic digester	0	1	0	9
Use dewatering technology	9	37	1	15
Control manure dust	7	32	0	8
Manage feed for nutrients	4	9	4	39
Raise heifers offsite	15	16	10	25

[1] A CNMP is a comprehensive nutrient management plan.

Source: 2005 Agricultural Resource Management Survey, version 4 (Dairy). Western dairy States in ARMS 2005 include AZ, CA, ID, NM, OR, TX, and WA. Traditional States are IA, IL, IN, ME, MI, MN, MO, NY, OH, PA, VT, and WI.

In traditional dairy States, only about 5 percent of dairy farms with fewer than 700 cows, and less than 30 percent of those with 700 or more cows, removed manure from the farm to other sites. Of those that removed manure, smaller farms removed 2 percent of the manure generated on the farm while large farms (700 or more cows) removed about 15 percent, on average. So as of 2005, manure removal was not widely used in traditional production regions.

Removal is far more important in the Western dairy States. Twenty-two percent of farms with fewer than 700 cows, and 57 percent of those with 700 or more, removed manure from the farm. Farms that remove manure also remove a lot more of it: the smaller farms removed 17 percent while the larger farms removed nearly half, on average, of the manure generated onfarm. Large Western dairy farms often raise no crops, whereas dairy farms in traditional dairy States usually grow crops for feed and retain substantial acreage for spreading manure.

About half of the manure removed from dairy farms was given away for free in each region, but Western dairies (large and small) paid fees to have over a third removed. Eastern dairies were able to sell over a third of the manure removed from those operations. Because Western dairies remove far more manure, the increased quantities likely depress manure values, such that fees for removal must be paid more often.

There were enough observations of manure removal on Western dairy farms to allow for estimates of the effect of manure sales or removal fees on farm revenues and costs, expressed in terms of dollars per cwt of milk produced for easier comparison. Among farms that sold manure, the median fee received was 6 cents per cwt. Reported payments for manure ranged from 5 cents at the 25th percentile to 6 cents at the 75th—a narrow range, and a very small contribution to revenues for those farms. For those who paid to remove manure, the median expense was 15 cents per cwt, and reported fees ranged from 7 cents at the 25th percentile to 30 cents at the 75th.

Reported removal expenses ranged more widely than revenues, but even at the high end the removal fees were modest compared with the production cost advantages reported for large operations. At the high end of 2005 manure removal fees (30 cents per cwt of milk at the 75th percentile, or 2.2 percent of production costs at a large farm), the expense is still a fraction of the production cost advantage of large operations.

Another manure management strategy is technology that allows for less costly, and therefore more distant, manure transport, and that can produce other benefits from manure. Large dairy farms were much more likely, in each region, to use technologies that ease the transport and promote the further processing of manure. Nevertheless, none of the technologies— manure separators, dewatering, or anaerobic digesters—is widely used (table 10). The anaerobic digester, which produces electricity from the methane in manure and leaves a dry, nutrient-intensive product that can be transported easily, appears on less than 9 percent of large farms in traditional dairy States, and is virtually absent elsewhere.

Farms can also reduce manure and nutrient production from a given herd by raising replacement heifers offsite, or by altering feed mixes to reduce manure or its nutrient content. Each of those strategies is common, although not widespread, on large dairy farms in traditional dairy States (table 10).

Little manure is currently removed in traditional dairy States. However, pressure is growing for a substantial expansion of manure removal because of the expansion of large dairy farms, the links between herd size and excess nutrient production, and the requirements in nutrient management plans. In turn, if the costs of manure removal were high, then regulations could, in principle, slow or reverse the expansion of large dairy farms. However, current evidence indicates that manure removal expenses add only modestly to milk production costs. Furthermore, given the abundant cropland in traditional dairy States, along with deep markets for gardening fertilizers, short hauling distances, and the opportunities for easier removal offered by available processing technologies, it seems likely that markets could well develop for manure removal as more large farms appear in traditional dairy States.

CONCLUSIONS

U.S. milk production is rapidly shifting to much larger dairy enterprises. The shift occurred first in Western dairy regions, and milk production has increased in Western States as more capacity has been added in large-scale farms. However, similar changes are occurring in more traditional dairy States, which are rapidly adding large operations and losing smaller ones.

Large dairy farms have substantial cost advantages over smaller farms, derived from the ability to take advantage of economies of scale. On average, farms with at least 1,000 cows realize costs, per hundredweight of milk produced, that are 15 percent lower than farms in the next largest size class (500–999 head) and 35 percent lower than farms with 100–199 head. Other evidence suggests that costs may continue to decline as herds increase to and above 3,000 head.

Smaller farms tend to get higher prices for their milk than larger farms. But cost differences tend to overwhelm this advantage: larger farms, especially those with more than 1,000 cows, are realizing economic profits while most smaller farms are realizing negative net returns. In turn, differences in returns are driving investment decisions that are shifting production to larger farms.

Still, some small farms realize economic profits—in that the value of their production exceeds their total costs, including operating costs, capital recovery costs, and the opportunity costs of the operators' time. Others, although not returning enough to encourage reinvestment, earn enough to remain in business.

The continued shift of production to larger operations will likely reduce industrywide costs, leading to lower dairy prices for consumers even as it forces more small operations out. But the shift also creates increased environmental risks associated with the concentration of manure-based nitrogen and phosphorus. Federal and State regulators have been applying rules to govern the storage, application, and disposal of nutrients. Some large specialized dairies already remove most or all of their manure to other sites, at modest costs; moreover, large dairies have begun to invest in technologies that allow for lower cost transport of manure-based nutrients. At present, the costs of manure treatment and transport are not large enough to offset the considerable production cost advantage held by large dairies. Given this, production should continue to shift to large dairy farms.

REFERENCES

Blayney, Don P. *The Changing Landscape of U.S. Milk Production.* U.S. Department of Agriculture, Economic Research Service. Statistical Bulletin No. 978. June 2002.

Chambers, Robert G. (1988). *Applied Production Analysis: A Dual Approach.* Cambridge, UK: Cambridge University Press.

Dao, James. "*In Ohio, One Farmer's Prosperity Is Another's Poison.*" New York Times, March 26, 2004.

El-Osta, Hisham, & Mary C. Ahearn. *Estimating the Opportunity Costs of Unpaid Farm Labor for U.S. Farm Operators.* U.S. Department of Agriculture, Economic Research Service. Technical Bulletin No. 1848. March 1996.

Gollehon, Noel, Caswell Margaret, Ribaudo Marc, Kellogg Robert, Lander Charles & Letson, David. *Confined Animal Production and Manure Nutrients.* U.S. Department of Agriculture, Economic Research Service. Agricultural Information Bulletin No. 771. June 2001.

Gullickson, Gil. "Success on Spruce Creek: A Multiyear Plan Springs a Successful Transition for this Dairy Farm." *Successful Farming*. Jan. 2006.

Henry, Fran. "*Big Farms, Big Problems?*" Cleveland Plain Dealer. Aug. 1, 2004.

Kellogg, Robert L., Lander, Charles H., Moffitt, David C. & Gollehon, Noel. "*Manure Nutrients Relative to the Capacity of Cropland and Pastureland To Assimilate Nutrients: Spatial and Temporal Trends for the United States.*" U.S. Department of Agriculture, Natural Resources Conservation Service. Publication No. nps00-0579. Dec. 2000.

Martin, Andrew. "*Wisconsin Dairy Feels Squeeze*," Chicago Tribune, Dec. 24, 2004.

Martin, Andrew. *"Dutch Hope U.S. a Cash Cow,"* Chicago Tribune, March 26, 2005.

McBride, William D. *Change in U.S. Livestock Production, 1969-92.* U.S. Department of Agriculture, Economic Research Service. Agricultural Economics Report No. 754. July 1997.

McBride, William D. & Catherine Greene. "*A Comparison of Conventional and Organic Milk Production Systems in the U.S.*" Paper presented at the annual meeting of the American Agricultural Economics Association, Portland, OR, July 2007.

Miller, James & Blayney, Don P. *Dairy Backg rounder.* U.S. Department of Agriculture, Economic Research Service. Outlook Report No. LDPM14501. July 2006.

Mosheim, Roberto & Knox Lovell, C. A. "Economic Efficiency, Structure and Scale Economies in the U.S. Dairy Sector." Paper presented at the annual meetings of the American Economic Association, Long Beach, CA, July 2006.

Panzar, John C. in [Richard Schmalensee and Robert D. Willig, eds.] *Handbook of Industrial Organization, v.1*, "Technological Determinants of Firm and Industry Structure", New York: North Holland Publishing Company, 1989.

Ribaudo Marc, Gollehon Noel, Aillery Marcel, Kaplan Jonathan, Robert Johansson, Jean Agapoff, Christensen Lee, Breneman Vince & Peters Mark. *Manure Management for Water Quality: Costs to Animal Feeding Operations of Applying Manure Nutrients to Land.* U.S. Department of Agriculture, Economic Research Service. Agricultural Economic Report No. 8244. June 2003.

Short, Sara D. *Characteristics and Production Costs of U.S. Dairy Operations.* U.S. Department of Agriculture, Economic Research Service. Statistical Bulletin No. SB974-6. Feb. 2004.

Stigler, George. "The Economies of Scale," *Journal of Law and Economics* 1 (Oct. 1958): 54-71.

Sumner, Daniel A. & Wolf, Christopher A. "Diversification, Vertical Integration, and the Regional Pattern of Dairy Farm Size," *Review of Agricultural Economics* 24 (Fall/Winter 2002): 442-457.

Tauer, Loren W. & Mishra, Ashok K. "Can the Small Dairy Farm Remain Competitive in U.S. Agriculture?" *Food Policy* 31 (October 2006): 458-468.

U.S. Department of Agriculture. *Economic Effects of U.S. Dairy Policy and Alternative Approaches to Dairy Policy.* Report to Congress. July 2004.

West, Lindsay. "New Mexico Dairy Finds Success in Numbers," *High Plains Journal, May 6, 2006.*

APPENDIX: DATA SOURCES

Dairy Farm Numbers, Size and Location

USDA's National Agricultural Statistics Service (NASS) publishes annual data on the size structure of dairy farms in *Farms, Land in Farms, and Livestock Operations* (before 2004, in February issues of *Milk Production*). NASS reports the number of operations, cow inventory, and milk production, by herd size class. Six classes are identified for State-level data—1-29, 30–49, 50–99, 100–199, 200–499, and 500 or more milk cows, with the largest class replaced by classes of 500-999, 1,000–1,999, and 2,000 or more for the nationwide data. The data are based on surveys of all large producers and samples of smaller producers.

The annual survey data are supplemented with the 1992, 1997, and 2002 Censuses of Agriculture. Confidential individual census records are used to analyze changes in farm size among large dairy farms. Census records report farm location and milk cow inventories at the end of the year, allowing us to summarize changes in the size distribution of dairy farms between census years. With an ongoing expansion of large commercial dairies, the 2002 census end-of-year estimates of large dairies will exceed the 2002 estimates in *Farms, Land in Farms, and Livestock Operations*, which rely on beginning and midyear surveys.

Data for Cost Analyses

Cost analyses are based on data from the dairy version of the 2000 and 2005 Agricultural Resource Management Survey (ARMS), a complex annual survey applied to a stratified random sample of all U.S. farms. ARMS is USDA's primary source of information on the financial condition, production practices, resource use, and economic well-being of U.S. farms and farm households.

ARMS is a large multiphase and multiversion survey. The sample is screened for continued operation and commodity coverage in Phase I, conducted in the summer of the reference year. In the following fall, randomly selected Phase I farms are surveyed in Phase II concerning their crop production practices and chemical use at the field or production unit level. Phase III, initiated during the winter following the reference year, draws information on farm income and expenditures, farm financial transactions, and the farm operator household.

Several versions of the Phase III survey are distributed. A short "core" version is distributed and returned by mail, while other versions require personal interviews with trained enumerators. One enumerated version (version 1) covers farms of all types. This version is used for the wage analysis reported in table 6 because it elicits information on off-farm hours worked, off-farm earnings, age, and education for farm operators and spouses. Other enumerated versions include sections aimed at specific commodity enterprises, including dairy versions covering the years 2000 and 2005. The 2000 ARMS dairy version covered 22 States: AZ, CA, FL, A, ID, IA, IL, IN, KY, MI, MN, MO, NM, NY, OH, PA, TN, TX, VT, VA, A, and WI. The 2005 version covered those States, plus ME and OR. The dairy versions elicit information on dairy enterprise production, inventories, expenses, assets, and technology use, as well as manure management and marketing practices. Further information on ARMS, including downloadable questionnaires, can be found at www.ers.usda.gov/Briefing/ARMS/.

Excess nutrients and manure management decisions

Data for the excess nutrient estimation were developed from ARMS, Phase III, version 1, which contains farm-level data on livestock inventories and crop acreage. The analysis focused on dairy farms in traditional production regions, where manure remains on the farm. Version 1, directed annually at all farms, covers several hundred dairy farms in each year, and a large sample could be obtained by drawing records for dairy farms in 11 States (IA, IL, IN, MI, MN, MO, NY, OH, PA, VT, and WI) from the 7 version 1 surveys carried out during 1996-2002. ARMS livestock inventory data were combined with external estimates of species-specific manure and nutrient production rates to generate estimates of manure-based N and P production on the farm. ARMS data on crop acreage were combined with external data on crop-specific nutrient uptakes to generate estimates of total nutrient uptake on the farm.

The data on manure management practices were drawn from the 2005 ARMS dairy version, which contained specific questions on those practices. The 2005 survey covered dairy farms in 24 States (listed above).

End Notes

[1] Dairy enterprises have calves and heifers, which are not yet ready to give milk; they may have bulls; and they have milk cows, which have given birth to calves. At any time, some fraction of a farm's milk cows are dry, usually in preparation for calving. Unless otherwise noted, herd sizes in this report refer to the number of milk cows on a farm, including dry milk cows.

[2] The largest class in State-level data covers farms with 500 or more cows.

[3] See Dao (2005), Henry (2004), and Martin (2005) for articles on the construction of large dairies in the Indiana-Michigan-Ohio area; Martin (2004) for Wisconsin; or Gullickson (2006) for Pennsylvania.

[4] The average cost curve represents how costs vary with output, for a given set of input prices. Changes in input prices, or technological innovations, could shift the position and shape of the curve.

[5] Technically, Chambers (1988) reserves the term "economies of scale" for a specific technological relationship—the increase in output attendant upon an equiproportionate increase in all inputs. However, if production is not homothetic, then that technological relationship will not capture the full change in longrun average costs attendant upon output growth (in non-homothetic production, cost minimizing factor proportions vary with output). Chambers uses the term "economies of size" for relationship between costs and output, but we use "economies of scale" to be consistent with the industrial organization literature (Panzar, 1989).

[6] McBride and Greene (2007) provide an analysis of organic dairy costs of production, and a comparison to conventional production, using the 2005 survey.

[7] The costs shown here, drawn from the dairy COP estimates at the ERS website, exceed those reported in Short (2004), who also used the 2000 ARMS dairy version. Short excluded several implicit costs—those associated with unpaid labor provided by the operator and the operator's family, farm overhead, and the value of the enterprise's land that is used to support the dairy enterprise (she includes implicit capital costs). Including these implicit costs raises the cost estimates more for smaller than for larger enterprises, so the scale economies apparent in table 3 are much larger than those in Short (2004).

[8] Feed accounts for large shares of total costs at dairy farms, ranging from 30 percent of total costs in the smallest class to 55 percent in the largest. Increases in feed prices, such as those resulting from increased ethanol-based demand for corn, have substantial effects on costs. Hired labor, often from Mexico and Central America, accounts for 10–12 percent of total costs at larger farms.

[9] However, if capital is not replaced, then the costs of maintaining it are likely to increase over time, leading to higher operating costs.

[10] Respondents could choose among six answers: less than a year, and 1, 2–5, 6–10, 11–19, and more than 20 years. The question emphasizes the operation instead of the operator, so transfer of the operation from father to son would not elicit a shutdown response.

[11] Cow inventories among farms with less than 100 cows actually fell by 26 percent between 2000 and 2005. That number reflects exit by farms operating in 2000 and entry by new farms.

[12] The Government also sets a support price for milk. However, the support price has generally remained below market prices in recent years and, since it applies to all producers, it does not favor smaller operators. See Miller and Blayney (2006) and U.S. Department of Agriculture (2004) for analyses of dairy pricing and policy.

[13] The analysis also accounts for the decision to work off the farm, thus taking account of the additional information that can be obtained by including those who do not work off the farm.

[14] Some unpaid labor hours are provided by family members who are under 16. ERS values the opportunity cost of their labor at the minimum wage in their State.

[15] The data in table 7 are drawn from censuses of agriculture, instead of the annual USDA/NASS surveys that underlie table 1. The census, conducted every 5 years, is less timely but more comprehensive.

[16] The very largest dairy farms may encompass several "pods" with barns and a milking facility for 2,000–3,000 cows in each.

[17] How big are the largest dairy farms? Eight farms in the 2002 census had over 10,000 milk cows, compared with 4 in 1997 and 1 in 1992. A farm in Indiana advertises that it has (in 2007) 30,000 milk cows in 10 pods on its site.

[18] Data are drawn from the 2005 ARMS dairy version, and show the ratio of milk cows to farmland operated and cropland across farms in each size class. For this figure, we chose size classes in accordance with regulatory standards.

[19] This analysis relies on data drawn from seven annual ARMS surveys conducted during 1996–2002. The version underlying these surveys carries less dairy farm detail than the dairy version, but nevertheless sufficient detail for these purposes, and a large sample size.

[20] Kellogg et al. report their estimates for animal units (AU), defined as 1,000 pounds of livestock (a mature dairy cow would be about 1.35 animal units, while a 250-pound hog would be 0.25 animal units). We converted their AU estimates to per-animal estimates.

[21] In generating these estimates, we assume that soybean and alfalfa plants are net fixers of N, with each crop fixing nitrogen (35 pounds per bushel of soybeans, and 135 pounds for alfalfa hay) into the soil.

[22] The history leading up to the 2003 rule is described in "National Pollutant Discharge Elimination System Permit Regulation and Effluent Limitation Guidelines and Standards for Concentrated Animal Feeding Operations (CAFOs); Final Rule," 40 CFR Parts 9, 122, 123, and 412, at http://www.epa. gov/fedrgstr/EPA-WATER/2003/February/Day-12/w3074.pdf .

[23] Described in the 2003 rule, p. 7226.

[24] Costs were sensitive to variations in the willingness of nearby crop farmers to accept manure. As willingness increased from 20 percent, costs would fall. If only 10 percent would be willing to accept manure, then estimated costs for large operations could double, still a small impact compared with the production cost advantages of size.

In: Transformation of U.S. Animal Agriculture
Editor: Justin M. Daigle

ISBN: 978-1-60876-938-4

Chapter 5

THE ECONOMIC ORGANIZATION OF U.S. BROILER PRODUCTION

James M. MacDonald
U.S. Dept. of Agriculture (USDA)

ABSTRACT

Broiler production in the United States is coordinated almost entirely through systems of production contracts, in which a grower's compensation is based, in part, on how the grower's performance compares with that of other growers. The industry is undergoing a gradual structural change as production shifts to larger broiler enterprises that provide larger shares of an operator's household income. Larger enterprises require substantially larger investments in broiler housing, and new or retrofi tted houses are also an important source of productivity growth in the industry. This report, based on a large and representative survey of broiler operations, describes the industry's organization, housing features, contract design, fees and enterprise cost structures, and farm and household finances.

Keywords. broilers, chickens, production contracts, broiler grower financial performance, chicken housing, chicken litter, poultry

ACKNOWLEDGMENTS

Special thanks for helpful comments and advice from those in government, academia, and industry, including Bill McBride, Dave Harvey, Bob Hoppe, Penni Korb, Joe Moore, Gary McBryde, John Rollins, Milt Madison and Dania Ferguson from the U.S. Department of Agriculture; Tom Vukina of North Carolina State University; H.L. Goodwin of the University of Arkansas; Dan Cunningham of the University of Georgia; Doug Parker of the University of Maryland; Bill Roenigk of the National Chicken Council; Kenny Bounds of MidAtlantic Farm Credit; and members of the Growers Committee of Delmarva Poultry Industries (DPI).

Angela Anderson and Cynthia A. Ray of ERS provided editing and design support for the manuscript.

SUMMARY

U.S. consumption of chicken averaged 86 pounds per person in 2006, more than triple the 1960 level. Chicken became a preferred option as incomes increased and chicken prices remained inexpensive compared with other meats, and as processors created new chicken products that appealed to consumer tastes. Increased consumption required increased production of broilers, which the industry accomplished through a tightly integrated production system that links chicken companies, called integrators, with independent broiler growers through production contracts. The industry's distinctive organization has contributed to its commercial success, but it now faces a series of challenges.

What Is the Issue?

After decades of rapid expansion, growth in both broiler production and productivity began to slow in the mid-1990s. Slowing growth creates challenges for industry decisionmakers, as they consider how to encourage further investments in capacity and new technology, and attempt to manage existing and aging production networks.

The broiler industry has a unique organization. Firms called integrators own processing plants, hatcheries, and feed mills, and contract with independent "grow- out" operations to raise their broilers to market weight. The contractual relationship between farmers and integrators, however, is coming under growing scrutiny from Congress and regulatory agencies, and the industry relies heavily on a particular kind of production contract that has attracted considerable attention.

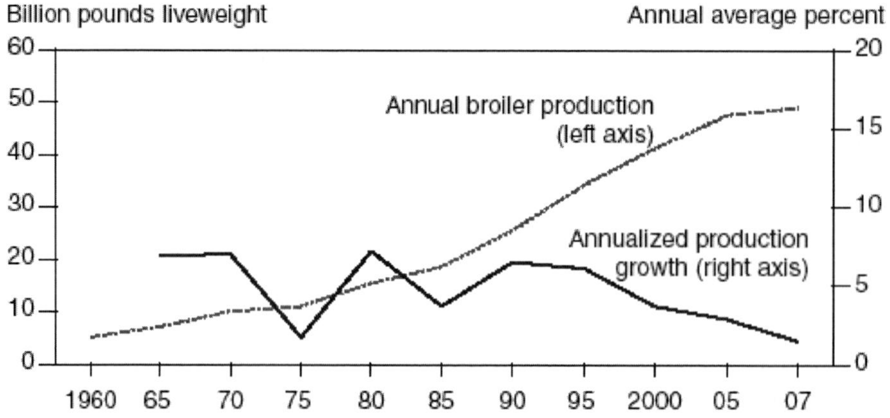

Sources: USDA, National Agricultural Statistics Service

Broiler production and production growth, 1960-2007

The broiler industry plays an important role in several public policy issues:

- Large animal feeding operations, including those raising broilers, are under increasingly strict environmental regulation by all levels of government; and
- The broiler industry has dealt with poultry diseases and associated biosecurity issues for many years, while growing public awareness of such threats plays an increasingly important role in industry and public policy planning.

What Did the Study Find?

Other industries use production contracts, but the broiler industry is distinguished by the dominance of such contracts and the methods by which growers are paid. Almost all broiler growers' contracts base the compensation on how each grower's performance compares with that of others.

Beyond that feature, however, contracts are far from uniform. Contracts can include other terms that tie base payments to actions that affect grower costs or that assign some expense or revenue categories to either the grower or the integrator. Contracts also cover a wide range of specifi ed durations, from just over a month to 15 years. Variations in contract design likely follow from differences in grower location, size, and type of broiler housing, but the wide variation in terms and payments makes it diffi cult for growers to evaluate contracts.

The industry's organization has contributed to its commercial success. High rates of productivity growth, along with new product innovation, led to high growth rates for chicken production, domestic consumption, and exports. Growth in production slowed noticeably, however, after the mid-1990s. With slower production growth, investment in new housing also slowed. New housing embodies new technology, so slowing investment can hinder future productivity growth, unless older houses can be effectively retrofi tted with newer equipment. Integrators are requiring such retrofi tting for some operations as a condition of extending their contract. For newer and larger operations, integrators are offering contracts of longer duration to induce them to continue to invest in new technology.

Broiler production is gradually shifting to larger operations, a trend common to most agricultural commodities. For operators of small broiler enterprises, off-farm income is the primary source of the household's income, and the broiler enterprise provides a modest amount of additional income. For larger operations, the broiler enterprise typically is the primary source of the household's income. As a result, operators of larger enterprises may be more sensitive to the income risks arising from energy price fluctuations and contract settlements. Contract features may need to be redesigned to adjust for differing risk exposures faced by growers.

Larger operations may realize scale economies in production, but they also concentrate poultry litter in localized areas. Litter is bedding material, such as wood shavings, sawdust, or straw, that is spread on the floors of broiler houses. When it is removed, it consists mostly of poultry manure, along with the original bedding, feathers, and spilled feed. In 2006, about 40 percent of used litter was spread on the farm's fields, while the rest was removed from broiler operations for field application elsewhere or for processing. There was enough of a market for litter in 2006 to allow growers to sell about a third of the litter removed from farms, but farmers had to give away the rest or pay to have it removed. Litter disposal remains a major issue confronting the industry.

How Was the Study Conducted?

The analysis relies on data drawn from a large-scale representative survey of producers, conducted as part of the annual Agricultural Resource Management Survey (ARMS), which is the U.S. Department of Agriculture's primary source of information on the financial condition of farm businesses and households and farm production practices.

Two ARMS versions collect financial and production information for all types of farms, but other versions target specifi c commodities and collect additional information on production practices, financial performance, and contractual relationships for those commodity enterprises. ARMS included a broiler version for the fi rst time in the survey conducted early in 2007, with a focus on performance during 2006. The survey's target population consisted of all operations that produced broilers for meat and had at least 1,000 broilers onsite at any time during 2006, in the 17 States that accounted for 94 percent of U.S. broiler production. Analyses in this report are based on responses received from 1,568 operations, out of 2,100 originally selected for the survey.

INTRODUCTION

This report describes the organization and economics of broiler production in 2006. Broilers, young chickens bred for meat, account for nearly all chicken meat and most poultry meat produced in the United States.[1] The broiler industry's processes are tightly controlled by firms called integrators, who operate processing plants, feed mills, and hatcheries, and who contract with farmers to grow broiler chicks to market weight. This report focuses on farms in the production or grow-out stage and their commercial relationship with integrators.[2]

Broiler production grew rapidly for many years, fueled by powerful shifts in consumption patterns. In 1960, Americans consumed about 28 pounds of chicken per person, compared to nearly 60 pounds of pork and 65 pounds of beef (Figure 1). Chicken consumption rose steadily in the ensuing decades, reaching 87 pounds per person in 2006, while per capita beef consumption remained virtually unchanged from its 1960 value and pork consumption declined. Increasing chicken consumption refl ected changes in consumer preferences, the introduction of many new chicken-based retail products, and declining relative prices for chicken.

Between 1960 and 2006, poultry prices rose by 2.7 percent per year while prices for all food rose by 4.2 percent per year, according to the Consumer Price Index (CPI). As a result, retail poultry became cheaper compared with other foods, encouraging a shift to poultry. Other price data indicate that rapid productivity growth was at the heart of poultry's superior price performance. Prices for prepared feeds, the major input used by poultry producers, rose by 2.5 percent per year during the same period, but wholesale broiler prices rose by just 1.3 percent per year as the industry found ways to increase feed efficiency.[3]

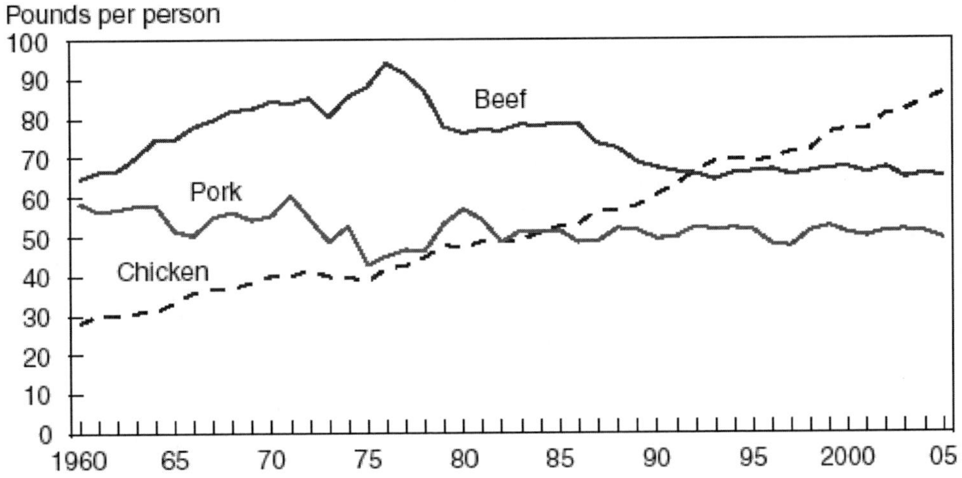

Source: ERS Food Availability (Per Capita) Data System, at www.ers.usda.gov/Data/Food Consumption.

Figure 1. Trends in per capita consumption, 1960-2005

Table 1. Growth in chicken production, 1960-2005

Item	1960-70	1970-80	1980-90	1990-2000	2000-05
		Average annual percentage growth in pounds produced (liveweight)			
Production	5.0	3.4	4.9	4.3	2.9
Contributions from growth in:					
Exports	-0.1	0.5	0.4	1.4	0.1
Population	1.3	1.2	1.2	1.1	0.9
Population	1.3	1.2	1.2	1.1	0.9
Per capita consumption	3.8	1.7	3.3	1.9	1.9

Source: ERS Food Availability (Per Capita) Data System, at www.ers.usda.gov/Data/FoodConsumption.

Per capita consumption increases of 2-4 percent per year, when combined with U.S. population growth of about 1 percent per year, provided the industry with substantial increases in demand over a long period. International markets provided a third source of demand growth, with increased chicken exports accounting for about a third of the total growth in production during the 1990s. Table 1 divides the growth in production into three parts—expanded exports, population, and consumption—and shows how changes in each affected the industry's production growth.

Overall production growth slowed sharply after 2000. While per capita consumption continued to grow at 1.9 percent per year, the contribution of population growth fell in comparison to earlier years and export growth fell sharply (table 1). Changes in exchange rates can affect the competitiveness of U.S. exports, and growth in foreign incomes can be an important force. In the near future, a declining dollar, combined with income growth in Asia, may boost export growth again.

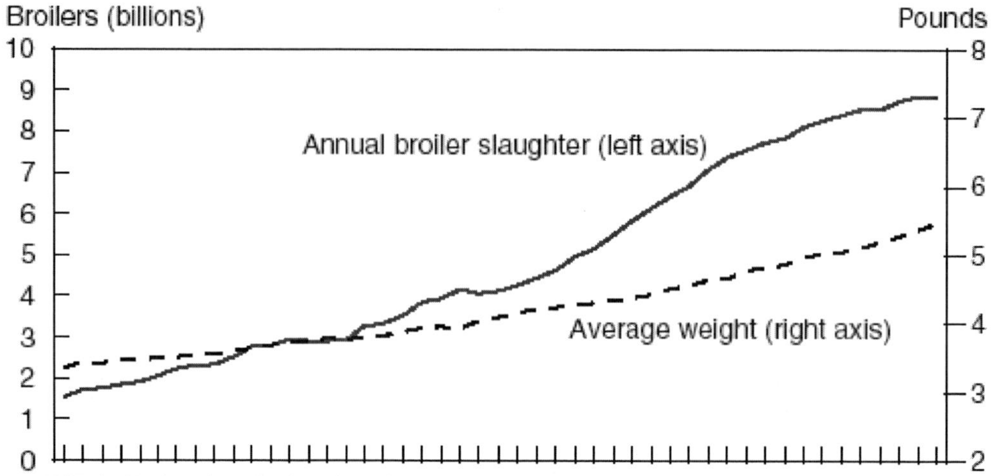

Source: USDA, National Agricultural Statistics Service.

Figure 2. Growth in broiler production, 1960-2006

Increases in total liveweight production have been realized from a rapid expansion of the total number of broilers produced and also from increasing the size of birds (Figure 2). Annual slaughter volumes rose from 1.53 billion birds in 1960 to 8.84 billion in 2006, while average weight rose from 3.4 to 5.5 pounds. In turn, increased production meant spreading to new locations, attracting new growers in existing locations, and increasing the size of production facilities. Slowing production growth is also apparent in figure 2. Before 1995, annual slaughter volumes of birds grew at a 4.6 percent annual rate, but at 1.7 percent afterward, even as average weight continued to increase.

Broiler production is organized in a distinctive manner. Most farms are linked to an integrator through a production contract, under which the integrator provides chicks, feed, veterinary services, and other inputs to the farmer, who grows the birds to market weight. Besides providing their own labor, farmers invest in specialized poultry housing (along with associated equipment), pay for any hired labor, and bear some or all of the cost of utilities. Because broiler housing is specialized and long-lived, the decision to produce broilers is a long-term commitment, and most producers have worked with their integrator for at least 10 years.

Integrators usually own hatcheries, feed mills, slaughter plants, and further processing plants—that is, they may be vertically integrated into all stages except for broiler production, where they rely on networks of growers assembled through production contracts. Integrators also contract with, or own, primary breeder companies that develop poultry breeding stock, and they contract with other farm operations to produce broiler eggs for hatcheries.

Production contracts are used in turkey and egg production and have been adopted widely in recent years throughout the hog industry. They can also be found in some areas of crop and horticultural production. But production contracts cover almost all broiler production, while other commodities also use a mix of other arrangements.

Production contracts in all commodities pay farmers for their growing services, not for the commodity. But the broiler industry uses a distinctive compensation arrangement. Farmers are most often paid on the basis of their relative performance, compared with other producers who deliver broilers to the integrator within a specifi ed time period, usually a week. Under a relative performance standard, all producers receive a base fee, but those who deliver more poultry meat for the number of chicks placed receive higher payments; differences in relative performance, therefore, are driven by differences in chick mortality and feed effi ciency. In turn, this type of contract shields producers from the common risks due to weather or input price fluctuations that affect producers as a group, since payment depends on relative performance and not on the average performance of the group (Knoeber, 1989).

Broiler production is geographically concentrated. Figure 3, drawn from the 2002 Census of Agriculture, shows where broilers are produced. The map clearly shows the concentration of production in Southeastern States and also shows how production is concentrated into dense networks within those States. Geographic concentration is driven by economies of scale in broiler production and slaughter, which encourage the growth of large facilities, and from the reductions in transportation costs for chicks, feed, and birds that can be achieved by locating processing plants, hatcheries, feed mills, and grow-out farms near one another. Pressures for geographic concentration are counter-balanced by the associated concentration of poultry litter, which can, in high concentration, create pollution risks to water and air resources, and by the risks from the contagious spread of poultry diseases within a network.

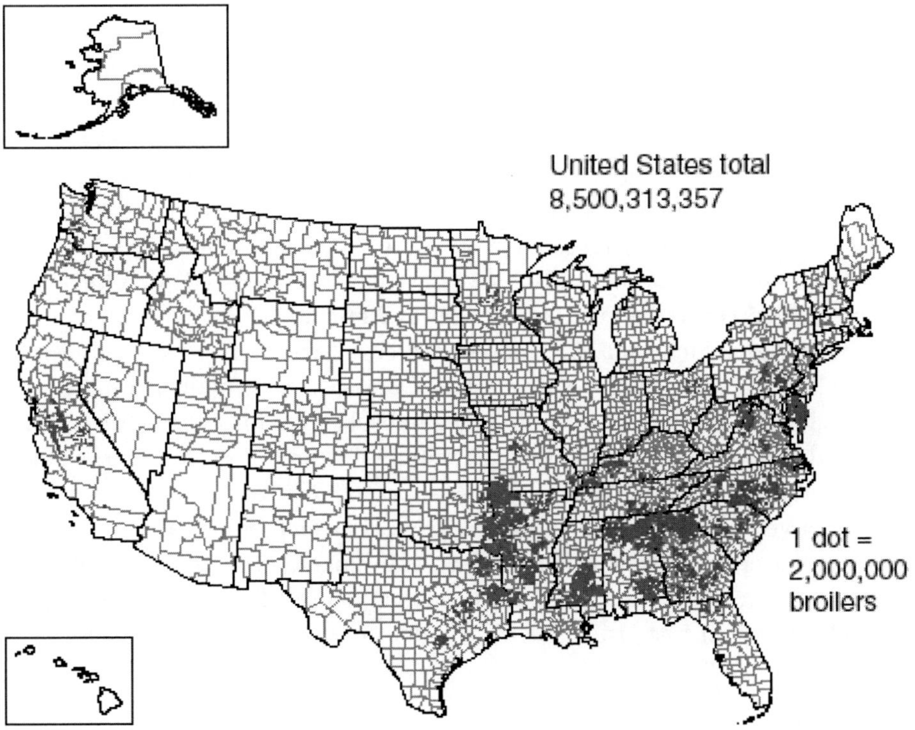

Source: USDA, National Agricultural Statistics Service.

Figure 3. Number of broilers and other meat-type chickens sold, 2002

Challenges Facing the Industry

The industry's distinctive organization has been an important contributing factor to its productivity and output growth (Rogers, 1979; Lasley, Henson, and Jones, 1983; Knoeber, 1989). But the industry now faces several challenges. The post-1995 slowdown in production growth is matched by a similar slowdown in productivity growth, resulting in retail price changes for chicken that now look much more like price changes for all food items.[4] Slowing productivity growth will press profi ts and lead to searches for different organizational designs.

While the industry's geographic concentration enables integrators to realize scale economies in production and processing, it also creates several risks and public policy challenges. For example, the worldwide spread of contagious poultry diseases is a growing concern and an important driver of industry biosecurity safeguards. But government agencies will also be involved in planning for and reacting to any serious outbreaks, and they need a deeper understanding of the industry's organization in order to carry out their responsibilities.

Because production occurs in localized networks, growers in most areas have very few integrators from which to choose. Many growers have only a single integrator in their area and most have no more than three. The lack of alternatives has led to controversy over production contracts and to legislative and regulatory proposals to regulate them.

Finally, high geographic concentrations of broiler litter can create increased risks of water and air pollution from excessive applications of the nitrogen and phosphorous in litter. Concerns over those risks have led to expanded regulatory intervention, expanded funding under USDA conservation programs, and public and private sector searches for solutions.

This report is designed to provide an overview of the industry's organization and economics, so as to provide objective and reliable background information. Our data are drawn from a large-scale representative survey of producers conducted early in 2007. The survey is part of the Agricultural Resource Management Survey (ARMS), an annual survey of U.S. farms that serves as the U.S. Department of Agriculture's primary source of information on the financial conditions and production practices of farm businesses and the well-being of farm households. ARMS has several versions. Two—a short version distributed by mail and a personally enumerated longer version—focus on all types of farms. Broiler operations do appear in those versions and some prior Economic Research Service (ERS) studies have used that data (Perry, Banker, and Green, 1999; MacDonald and Korb, 2008).

Other survey versions (up to three each year), which are also personally enumerated, focus on producers of selected commodities and include detailed questions on commodity production practices, as well as contractual and marketing relationships. A broiler version was included, for the fi rst time, in an ARMS conducted in early 2007; the questions focused on performance during 2006.

Conducting a Broiler Version of the ARMS

The survey focused on commercial producers of broilers grown for meat. Therefore, a sample was drawn from a target population consisting of all operations that produced broilers for meat and had at least 1,000 broilers onsite at any time during 2006. The sample defi nition excluded operations that raise broilers for show or private consumption, as well as egg-laying, hatchery, and broiler breeder operations. To conduct the survey effi ciently, standard practice for commodity-specifi c ARMS versions is to limit the sample to major production States—in this case, 17 States that account for 94 percent of U.S. broiler production.[5]

ARMS samples are randomly selected from a list of operations compiled from the Census of Agriculture, commercial databases, and USDA program databases. In June 2006, sample farms were screened for continued operation during ARMS survey Phase I. The questionnaire was finalized in November 2006, and survey enumerators conducted their interviews in February and March 2007. Of 2,100 operations targeted for interviews for the broiler version, 1,602 usable survey responses were received. Some respondents were still in farming, but did not produce broilers for meat during 2006, leaving 1,568 broiler producers for analysis (a 75-percent response rate). Most of this report focuses on the 1,546 operations that reported having a production contract for broilers (22 were independents, processor-owned, or did not respond to the question).

To obtain more reliable estimates, some farms have a higher probability of sample selection. For example, larger operations are more likely to be selected for interviews than smaller, and selection probabilities also vary across geographic areas. Each sample farm represents a number of other farms from a similar geographic location and size class. These weights (the number of farms that each sample point represents) range from 3 to 40 farms. When sample observations are weighted to refl ect their varying selection probabilities, population estimates for production and other industry charac teristics can be generated.

How Production is Organized

There were an estimated 17,440 commercial broiler farms in the 17 surveyed States in 2006, with production of 8.44 billion broilers (table 2). Since total nationwide slaughter amounted to 8.84 billion broilers in 2006 (figure 2), the data from this 17-State sample represent 95 percent of total production.

Table 2. Broiler production in 2006, by type of operation

	All farms		Farms reporting broiler removals		
Type of operation	**Obs.**	**Farms**	**Obs.**	**Farms**	**Removals**
Production contract	1,546	17,200	1,543	17,183	8,310,308,738
Processor-owned	12	163	12	163	84,166,446
Independent	6	52	6	52	31,411,423
More than one type	2	14	2	14	8,219,932
Refusal/don't know	2	11	2	11	5,265,540
All operations	1,568	17,440	1,565	17,423	8,439,372,079

Notes: "Obs." refers to sample observations. Three sample farms with production contracts failed to report any birds removed, leaving 1,543 respondents in the removals column.

Source: 2006 Agricultural Resource Management Survey, version 4.

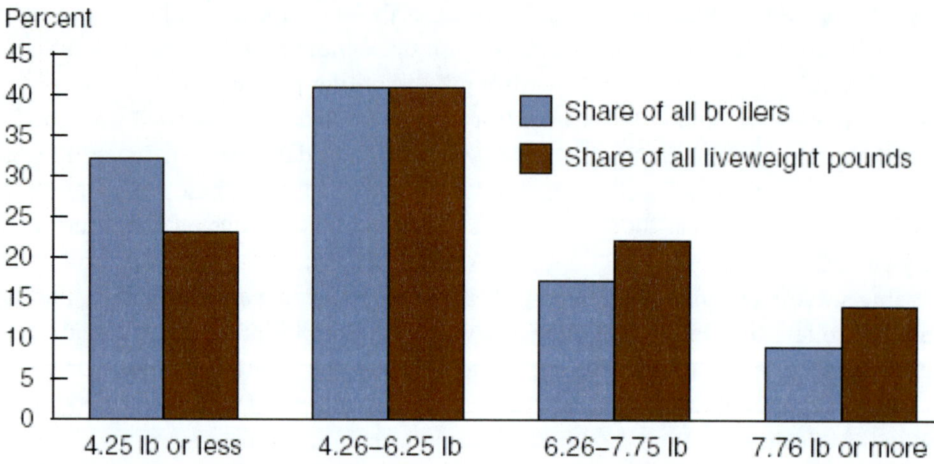

Source: 2006 Agricultural Resource Management Survey, version 4.

Figure 4. Broiler production in 2006, by size of bird

Production contracts dominate the industry, accounting for almost all farms and broilers (table 2). The survey did identify a few independent operations (0.4 percent of birds produced) and some operations that were owned by processors (about 1 percent of farms). Respondents were also asked whether they produced certifi ed organic broilers or free-range broilers ("free range" is operator-defi ned and not necessarily organic). About 1.7 percent of operations were certifi ed organic (1.4 percent of broilers), while a smaller fraction (0.44 percent of operations) reported that they produced free-range broilers.

Farms specialize according to the size of the bird produced; smaller broilers are used in the food service and restaurant trade, while the largest are cut into parts in processing plants.[6] The survey sorted production into four size classes according to weight at removal (figure 4). The most common, birds between 4.26 and 6.25 pounds, captured just over 40 percent of total production, measured by birds or by weight. The smallest class (4.25 lb. or less) accounted for 32 percent of all birds and 23 percent of all liveweight pounds produced in 2006, while the largest class accounted for 9 percent of birds and 14 percent of liveweight production. Over time, production has shifted to larger birds, refl ecting the growth of de-boned and further processed products. Large birds are grown in larger houses equipped with more modern climate controls. They also spend more time on the farm, and hence consume more labor, feed, utilities, and housing services, so producers of larger birds receive higher compensation per bird.[7]

Housing

Broiler houses are a major investment for growers. A single large house of nearly 30,000 square feet can cost $300,000, and most growers have multiple houses. Housing also plays an important role in the industry's productivity growth, as improvements in housing design and climate control systems can lead to improved feed effi ciency, lower bird mortality, and reduced costs of feeding, litter management, and bird removal.

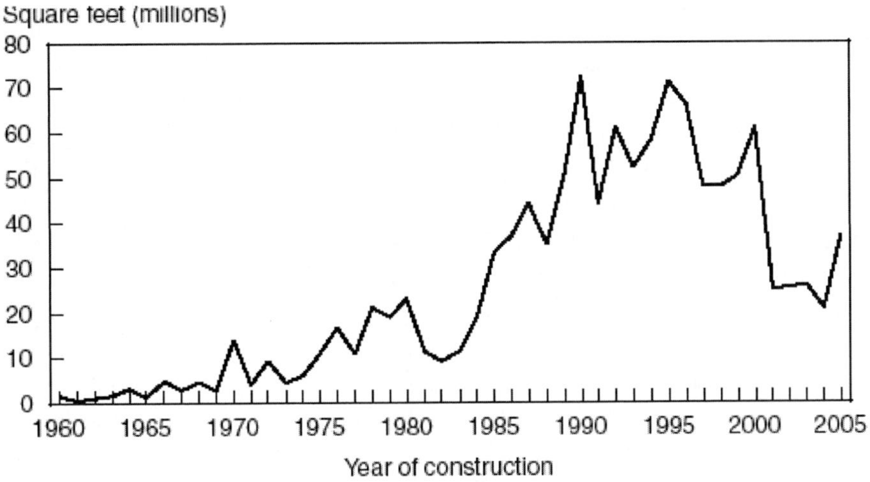

Source: 2006 Agricultural Resource Management Survey, version 4.

Figure 5. Broiler housing capacity, by year of construction

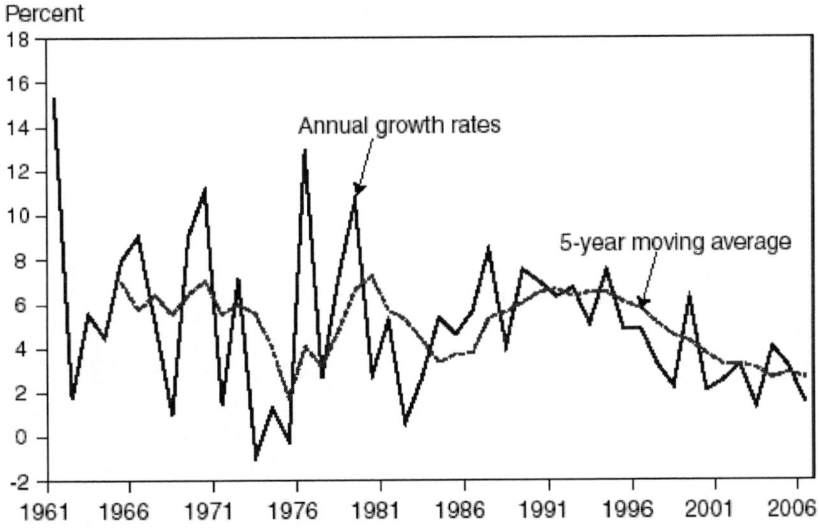

Source: USDA National Agricultural Statistics Service.

Figure 6. Annual growth in broiler slaughter (liveweight), 1961-2006

There were over 70,000 broiler houses in use for meat production in 2006. Their age structure follows a hump-shaped profi le (figure 5), with nearly two-thirds of capacity built in a 15-year period from 1986 through 2000. Investment dropped sharply after 2000, as considerably less capacity was built in the 6-year period from 2001 through 2006 than in any of the three preceding 5-year periods (1986-1990, 1991-95, and 1996-2000).

The slowdown in new house construction mirrors the slowdown in the industry's growth (Figure 6). Annual production of broilers grew by nearly 6 percent per year in the late 1980s and early 1990s, when measured by live- weight pounds at slaughter plants. Growth rates fell after the mid-1990s, to 2-3 percent per year in 2001-2006. As a result, far fewer new houses were needed to meet additions to production.[8]

The diffusion of technologies in the industry can be traced by sorting houses according to vintage—that is, the year in which they were built (table 3). Houses have become steadily larger over time. Whereas the typical house built in the 1960s was about 12,750 square feet, recently built houses average over 20,000 square feet, and large houses built in 2005 and 2006 are much larger—up to 36,000 square feet. Most houses built before 2000 have side curtains—fabric that can be raised or lowered to help control the climate in a house with natural ventilation. While many recently built houses also have side curtains, they are more likely to have solid walls with equipment added for climate control.

Two important types of climate control equipment are tunnel ventilation and evaporative cooling cells. Tunnel ventilation systems consist of large fans at one end of a broiler house and air inlets at the other end. The fans pull air through the house, removing heat from the building and creating a wind chill that provides further cooling. Evaporative cooling systems can be activated when tunnel ventilation alone fails to provide suffi cient cooling. The systems are located on the outside of the house near air inlets. Perforated pads in the systems are moistened, either through the use of fogging nozzles that spray water or through pipes that seep water through the pads. The cooling pads then lower the temperature of the air as it is pulled through the pads and the house. Additional control equipment can be used to automatically monitor a house's temperature and humidity, adjusting the climate accordingly.

About 75 percent of broiler houses had cooling cells and tunnel ventilation in 2006, and newer houses are much more likely to have them (table 3). Over 90 percent of houses built after 2000 had those features, compared with 70 percent of houses from the late 1980s and 50 percent from the 1970s. Older houses that have cooling cells and tunnel ventilation have probably been retrofi tted, with the equipment installed after the houses were originally constructed.

Table 3. Broiler housing, by vintage

Vintage (year built)	Houses		Mean size (Sq. ft.)	Percent of houses with:		
	Number	Share of total (%)[1]		Side curtains	Cooling cells	Tunnel vent.
No year	1,013	1.4	14,466	66	29	57
pre-1960	345	0.5	12,340	70	50	36
1961-1965	619	0.9	12,751	80	28	11
1966-1970	1,292	1.8	12,754	87	44	40
1971-1975	2,883	4.1	13,110	82	49	53
1976-1980	5,362	7.6	14,694	72	55	53
1981-1985	4,486	6.4	16,449	68	64	64
1986-1990	12,291	17.4	16,259	75	71	70
1991-1995	16,303	23.1	17,678	74	78	79
1996-2000	15,142	21.5	18,765	75	85	88
2001-2006	10,861	15.4	20,180	48	92	94
All houses	70,597	100.0	17,352	70	75	76

[1] Column sums may not add to 100 due to rounding.

Source: 2006 Agricultural Resource Management Survey, version 4. Only includes houses for farms that have production contracts.

Table 4. Experience and broiler production

	Years operation has produced broilers			
Item	**Under 6**	**6-10**	**11-20**	**Over 20**
Farms	770	3,354	7,073	5,802
Share of broiler farms (%)	4.5	19.5	41.2	33.8
Share of broiler removals (%)	6.6	22.9	41.3	28.4
Mean operator age (years)	48	49	54	60
Financial characteristics				
Debt per sq ft of housing ($)	4.55	4.81	2.42	1.69
Share with no debt (%)	14	13	25	43
Housing characteristics				
Average age (years)	11	12	16	24
Tunnel ventilation (%)	88	85	76	55
Evaporative cooling (%)	91	80	72	57
Side curtains (%)	43	72	72	73
Contract duration				
Five years or more (%)	40	22	15	6
One year or less (%)	46	58	62	73

Notes: Shares of farms and pounds sum to 100 when refusals are added (1.0 percent of farms and 0.8 percent of pounds). Debt, per square foot of housing, is reported for all farms, not just those with debt.

Source: 2006 Agricultural Resource Management Survey, version 4, production contracts only.

Table 5. Production and the number of houses on broiler operations

	Birds removed	**Pounds removed**	**Capacity utilization**
Operations with:		**Means**	
Two houses	227,971	1,163,653	35.5
Three houses	340,298	1,775,921	36.1
Four houses	472,302	2,584,358	36.8
Six houses	722,530	4,079,423	38.7
Eight houses	896,440	5,229,069	36.9
All operations			
Mean	483,600	2,615,251	36.8
Median	402,500	2,211,600	36.5

Notes: Capacity utilization equals pounds removed per square foot of housing capacity. The table lists the fi ve most common classes, according to the number of houses.

Source: 2006 Agricultural Resource Management Survey, version 4, production contracts only.

Entry and Experience in Broiler Production

Slowing industry growth means fewer recent entrants to broiler production. The survey asked respondents for the number of years that the operation had been producing broilers.

Only 4.5 percent of farms, with 6.6 percent of production, produced broilers for 5 years or less (table 4). In contrast operations that, by 2006, had been in operation for 6-10 years accounted for 20 percent of farms and 23 percent of production. At the other end of the spec-trum, about a third of all operations had been in business for at least 20 years.

An operation's age has a strong connection to its housing and fi nances. Older operations are smaller with much older housing and technology. Many older operations have no debt, while those with debt carry much less, per square foot of capacity, than newer operations.

New operations, in business for 5 years or less, have houses that are 11 years old on average, indicating that many recent entrants bought out existing operations with older facilities. Just under half of the new operations in table 4 also had new houses. Those operations carried higher debt loads—$6.52 per square foot, on average, compared to $3.03 for new operations with older houses— and 99 percent of their houses had evaporative cooling and tunnel ventilation.

How Large Are Broiler Operations?

Among farms with production contracts, average production in 2006 was 483,600 birds and 2.65 million pounds removed (table 5). But because averages are infl uenced by a few very large operations, two other measures can also provide useful information on farm sizes.[9]

The typical operation, as measured by the median (half of operations produced more and half less), produced 402,500 birds and 2.2 million pounds. While the typical contract broiler farm produced 402,500 birds in 2006, the typical broiler came from an operation that produced about 605,000 birds (half of broiler production came from farms that produced more than 605,000 birds, and half came from farms with less). Using this last measure, production is shifting to larger operations over time. Hoppe et al. (2007) use Census of Agriculture data to show that this midpoint farm—where half of broiler production came from larger farms and half from smaller—was 520,000 broilers in 2002 and 300,000 in 1987, just about half the size of the midpoint broiler enterprise in 2006.

Table 6. Size distribution of broiler operations

Number of houses	Farms	Broilers removed	Pounds removed	Capacity (sq. ft.)
	Percent of total			
nr	0.5	0.2	0.2	0
1-2	27.3	11.6	10.7	11.0
3-4	43.1	38.0	37.4	38.0
5-6	18.7	25.4	26.0	25.0
7-8	6.1	10.9	11.3	11.8
9-10	1.7	4.2	4.2	4.2
11-12	1.2	3.4	3.6	3.5
13-18	1.6	6.4	6.7	6.6
All farms	100.0	100.0	100.0	100.0
	Million			
Totals	17,183	8,310	44,815	1,221

Notes: Columns may not add to 100 because of rounding. Some operations did not report the number of houses, or had none, and they are designated with an "nr."

Source: 2006 Agricultural Resource Management Survey, version 4, production contracts only.

We can also look at size according to the number of houses on an operation. Most are small: 70 percent had between 1 and 4 houses, and together they accounted for just under half of broiler production, while operations with 5 to 6 houses and operations with 7 or more houses each accounted for just over a quarter of production. The largest operations responding to the survey had 18 houses, although enterprises of that size are not common (table 6).

Although production is shifting to large operations, family farms still dominate. Integrators directly own farms that account for 1 percent of production. Among farms with production contracts, 88 percent are sole proprietorships (table 7). While corporations account for just over 10 percent of production, most are family corporations in which more than half of the stock is owned by people related by blood or marriage. Large family farms often seek to incorporate, and that pattern holds among broiler operations—35 percent of those with 13-18 houses are incorporated, compared with only 6 percent of those with 3-4 houses (table 8).

Table 7. Legal organization of contract broiler operations

Type of organization	Farms	Broilers
	Percent	
Individual(sole/family proprietorship)	87.9	83.6
Legal partnership	4.6	5.8
Family corporation[1]	5.2	7.2
Nonfamily corporation	2.2	3.2
Other[2]	0.1	0.2
Total	100.0	100.0

[1]Family corporations are those in which more than half of the voting stock is held by people related to one another by blood or marriage.

[2]Other includes estates, trusts, and cooperatives. Source: 2006 Agricultural Resource Management Survey, version 4, production contracts only.

Table 8. Farm organization, by size of broiler operation

Number of houses	Percent of farms that are incorporated	Share of gross cash income from broiler contract fees
1-2	4	68
3-4	6	88
5-6	8	87
7-8	15	87
9-10	10	88
11-12	28	79
13-18	35	75
All farms	7	85

Note: Gross cash farm income includes fees from production contracts, revenues from cash sales and marketing contracts, government payments, land rents received, and revenues from other farm activities, such as grazing, custom work, and machine hire.

Source: 2006 Agricultural Resource Management Survey, version 4, production contracts only.

Production Contracts

Broiler production is a long-term commitment—houses are expensive, built to last for many years, and have few alternative uses. Because of that, producers tend to have long-term relationships with their integrator. On average, survey respondents reported holding contracts for 13 years with their current integrator.

Actual contracts often specify very short durations (figure 7). Forty-fi ve percent of respondents reported that their contracts were flock-to-fl ock, explicitly covering only the birds currently in their houses (5-10 weeks).[10] But durations also range widely. Eight percent of production contracts specifi ed a term of at least 7 years, while the longest run 15 years.

Long durations tend to be offered to newer and larger operations (table 4). Forty percent of recent entrants had long-term contracts (a specifi ed duration of at least 5 years) compared with only 6 percent of operations in business for at least 20 years. Among recent entrants, those with long-term contracts averaged half again as much production as those with shorter contracts.

Contracts and Competing Integrators

Because production is so localized, most producers have few integrators to choose from. Nearly a quarter (24.7 percent) reported that only a single integrator served their area, while another 28.7 percent reported two and 21.7 percent reported three. Those findings may overstate the number of integrators farmers can choose among if some integrators are not actively seeking new growers. In the 2004 ARMS (version 1), we asked 545 respondents with broiler production contracts whether they had alternatives, in the form of other contractors, cash markets, or both, to their current contractor. Most (59 percent) responded that they had none—far more than the 2006 fraction reporting a single integrator in the area. The responses would be consistent with one another if some integrators, particularly those in areas with 2 or 3, were not recruiting growers.

Integrators will not necessarily be able to exercise market power, even if there are only one or two in an area, because the competition for growers depends on more than just the number of integrators in a market. Integrators must recruit growers away from other activities, such as producing other commodities on the farm or working off the farm. Those outside options for growers set limits on the degree to which integrators can impose low fees on growers. There is little empirical evidence, however, of the effects of integrator concentration on grower returns.

Once a contract has expired, growers may have to retrofi t their houses with new capital equipment in order to gain a contract extension. These expenditures can be substantial. Between 2004 and 2006, farms spent a total of $650 million on capital improvements to their broiler enterprises, an average of $38,000 per farm. That average is skewed by a few large expenditures, but one operation in six spent at least $50,000 and another one in eight spent between $25,000 and $50,000. Retrofi tting is related to the age of the operation' s facilities, and those with the oldest houses were much less likely to make signifi cant investments in new equipment (figure 8).

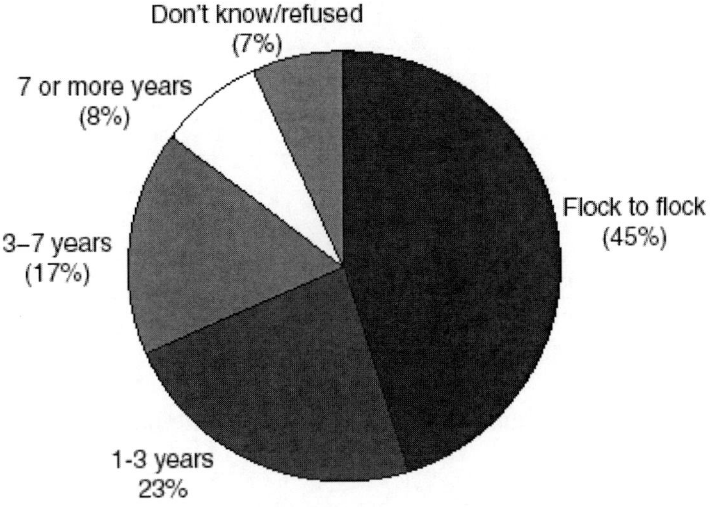

Source: 2006 Agricultural Resource Management Survey, version 4.

Figure 7. Contract duration among broiler operations

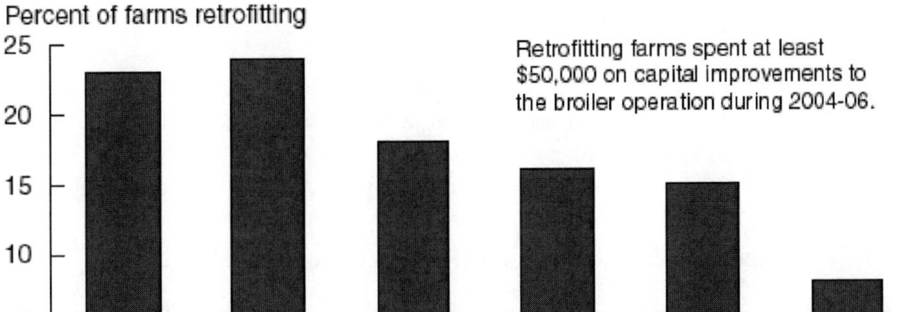

Source: 2006 Agricultural Resource Management Survey, version 4.

Figure 8. Farms with older houses are less likely to retrofit

Fees Paid in Production Contracts

Broiler farms are quite specialized and depend heavily on fees from broiler production contracts, which account for 85 percent of their gross cash farm income, on average (table 8).[11] In the hog industry, where production contracts are also used widely, contract fees average about 27 percent of gross cash income on farms with production contracts for market hogs.[12]

Table 9. Contract features related to compensation

Contract includes:	Percent of contracts with feature
Performance-based payment incentives	92.7
Pay for performance, compared with other growers	86.9
Adjustments for market price of broilers	13.2
Provisions for catastrophic payments	17.9
Seasonal adjustments for energy prices	55.7
Prices for energy purchased from specifi ed dealer	5.6
Facility fi nancing from the contractor	6.0

Source: 2006 Agricultural Resource Management Survey, version 4, production contracts only.

Table 10. Payments received under production contracts

Broiler size class	Mean fee per bird	Fees per pound		
		Mean	10th percentile	90th percentile
		Cents		
4.25 lb or less	20.4	5.3	4.1	6.6
4.26-6.25 lb	26.6	5.0	3.9	5.9
6.26-7.75 lb	34.3	5.0	3.6	6.1
7.76 lb or more	37.3	4.6	3.1	5.7
All birds	27.7	5.0	3.8	6.2

Note: Ten percent of fees in any size class are at or below the 10th percentile value, and 10 percent are at or above the 90th percentile value.

Source: 2006 Agricultural Resource Management Survey, version 4, production contracts only.

Table 11. Production practices specifi ed in contracts

	Percent responding:		
The contractor requires:	**Yes**	**No**	**Don't know**
Testing fl ocks for pathogens:			
Avian infl uenza	62.6	11.6	25.8
Salmonella	52.3	14.6	33.1
Other pathogens	49.4	14.3	36.3
Specifi ed production practices:			
Flocks must be all-in, all-out	84.8	5.9	9.3
Houses must be cleaned out after each fl ock	31.1	59.7	9.2
Operation must have HAACP plan[1]	54.3	16.0	29.7
Specifi ed animal welfare practices[2]	71.2	10.1	18.7
No antibiotics in feed or water[3]	42.4	27.2	29.4

[1] HACCP refers to Hazard Analysis and Critical Control Point, a systematic set of principles for improving food safety.

[2] The question specifi ed, as examples, standards for minimum space per bird and Humane Farm Animal Care (HFAC) certifi cation.
[3] The question contained the qualifi er "unless the birds are ill."
Source: 2006 Agricultural Resource Management Survey, version 4, production contracts only.

Broiler contracts specify compensation arrangements, which usually depend on the grower's performance in raising chicks to market weight, with performance measured in comparison to other growers (table 9). Growers who deliver more meat, for the amount of feed and chicks provided, get rewarded under the relative performance design of contract compensation. In turn, that depends on feed effi ciency and mortality. Other factors may infl uence contract payments. Some contracts adjust compensation for energy prices, market prices of broilers, or local catastrophes. Furthermore, pricing schedules may vary across locations, depending on the amount of competition that an integrator faces in attracting growers. But relative performance has a powerful effect on compensation, and the strongest performers can be paid 50 percent more per pound than the weakest.

Fees received under production contracts vary with the size of the bird produced (table 10). Operations that produce larger birds receive more per bird—37 cents, on average in the largest size class, compared with 20 cents in the smallest. But larger birds also cost growers more, because they have longer production cycles, thus using more energy, labor, and housing services. For that reason, fees are better analyzed on a per pound basis, where they average about 5 cents (liveweight), with average fees falling slightly as bird sizes increase. Because fees can vary widely with relative performance, the 10th and 90th percentile values for fees are also reported in table 10. This "interdecile range" captures the variation in fees around the average (10 percent of growers receive fees that are at or below the 10th percentile value, and 10 percent receive fees that are at or above the 90th percentile value). While average fees cluster around 5 cents per pound, fees at the 10th percentile can fall to 3 and 4 cents per pound (varying with bird size), while fees at the 90th percentile cluster around 6 cents per pound. Some of the variation may refl ect differences in relative performance among producers, but the variation can refl ect other elements of contracting as well. For example, integrators in some areas bear at least part of utility or litter expenses, and fees will be lower in those areas.

Contracts may require growers to carry out certain production practices, and some of those requirements may affect grower costs and fees. The survey asked about several practices that related to testing of flocks for avian infl uenza, salmonella, and other pathogens (table 11). Few respondents (11.6 to 14.3 percent) asserted that such testing was not required, but many reported that they did not know (25.8 to 36.3 percent). Because some tests may be conducted by the integrator, the "don't know" response is reasonable and informative. Over 40 percent of operations reported that growth-promoting antibiotics were not used in their feed; in this case as well, it is not surprising to see a high fraction of "don't knows" (29.4 percent), since feed is provided by the integrator.

Eight-fi ve percent of growers said that production was carried out on an all-in all-out basis—that is, all birds were removed at the same time and all chicks were placed at the same time. All-in-all-out production limits the spread of disease and also allows operations to tailor feed mixes to birds of a specifi c age. Over 70 percent reported that their contractor required them to follow specifi c animal welfare practices, such as space standards per bird, while only 10 percent reported that they were not required to follow such practices.

Two other practices were tied to food and poultry safety concerns. As of 2006, just over half of all operations were required to have a HACCP (Hazard Analysis and Critical Control Point) plan in place for food safety. And about a third of contracts required houses to be fully cleaned out, washed, sanitized, and dried after removal of each flock, while more than half did not.[13] The alternative, removing some or no litter after a fl ock is removed, with full clean-out occurring only after several flock removals, is less costly for the grower. Each approach seems to be widely used; about a quarter of respondents did not fully clean out any houses in 2006.

Expenses on Broiler Operations

The size of the operation affects the form of broiler enterprise expenses. For example, large and small operations provide labor in different ways (table 12). In the smallest enterprises (1-2 houses), the primary operator reports providing the broiler enterprise with 25 hours per week, on average. For those with 1-2 houses, a broiler enterprise provides additional income, but it is unlikely to be the primary source of employment. Operator hours rise steadily with enterprise size to 35 hours per week (3-4 houses) and to 45 hours (9 or more houses) with some additional hours provided by spouses and other operators.

Smaller operations use virtually no hired labor (table 12). In contrast, larger enterprises rely on considerable amounts of hired labor, which allows for greatly expanded production for a given time commitment by the operators. Among large enterprises, hired labor compensation amounts to 0.45 to 0.60 cents per pound or about 10 percent of average production contract fees.

Utility expenses are the major cash operating expense faced by most growers, and amount to 1.1-1.2 cents per pound, a signifi cant cost when production contract fees average about 5 cent per pound (table 13). Electricity expenses vary from 0.36 to 0.41 cents per pound of meat produced, with no apparent advantage to larger sizes. Most growers spend more on fuels with expenses ranging from 0.6 to 0.8 cent per pound, although the largest operations realized noticeably lower expenses in 2006.

Table 12. Labor commitments in contract broiler production

Number of houses	Weekly hours		Labor inputs to production	
	Primary operator	All operators	Unpaid hours	Paid labor compensation ($)
			Per 1,000 pounds produced	
1-2	25	30	1.72	0.00
2-4	35	40	0.97	0.09
5-6	40	49	0.67	1.01
7-8	40	50	0.53	2.44
9-10	45	45	0.41	2.65
11-12	45	50	0.35	4.42
13-18	45	45	0.27	5.97
All farms	32	40	0.96	0.15

Note: Estimates are median values in each size class (half of farms in a class have greater values and half have smaller).
Source: 2006 Agricultural Resource Management Survey, version 4, production contracts only.

Table 13. Utility expenses in contract broiler production

Houses	Fuels & oils	Electricity
	Cents per pound produced	
1-2	0.760	0.395
3-4	0.753	0.395
5-6	0.684	0.396
7-8	0.679	0.378
9-10	0.802	0.363
11-12	0.760	0.367
13-18	0.572	0.405
All farms	0.735	0.394

Note: Estimates are median values among operations in each size class.
Source: 2006 Agricultural Resource Management Survey, version 4, production contracts only.

Fuel expense can also be an important feature in contracts—three-quarters of the smallest producers received some fuel assistance from integrators in 2006, compared with 40 percent of the largest, with the incidence of assistance falling steadily as enterprise size increased. Assistance took two primary forms: reimbursement for fuel expenses or adjustment of compensation to refl ect seasonal changes in fuel prices. If energy prices continue to exhibit the wide variation seen in recent years, we can expect to see more such assistance, as well as efforts to redesign houses for greater energy effi ciency.

Litter Management Practices

Poultry litter is bedding material, such as wood shavings, sawdust, or straw, spread on broiler house fl oors. After being used, litter consists mostly of poultry manure, along with the original bedding, feathers, and spilled feed. The manure contains nutrients, including nitrogen, phosphorous, potassium, and calcium that can be used to fertilize cropland. Excessive applications of nutrients, however, can create environmental risks to water and air resources. Litter management, therefore, becomes an important issue.

Farms may apply used litter to cropland or they can remove it to other farms or other uses. Most farms (about 71 percent) removed at least some litter from the farm in 2006, and a little over half of those removed all of it (table 14). In the aggregate, 61 percent of all broiler manure produced on farms in 2006 was removed and used off the farm.

Thirty-six percent of the litter removed from farms was sold. Revenue from litter sales, about 0.2 cent per pound of liveweight production, therefore added about 4 percent to the operation's gross income for growers who could find litter buyers.

About 60 percent of operations applied litter to their own fields (table 14), and about half of those were able to field-apply all of their litter. Thirty- nine percent of all litter stayed on the farm and was applied to fields on the operation. For those that did so, survey responses indicated that about 1 acre of cropland was fertilized with litter for every 4,000 broilers produced (for average size birds). For an operation that removed 400,000 birds in a year, and aimed to field-apply all litter, 100 acres would therefore be required, on average.[14]

Table 14. Methods of managing litter

Methods of litter mangement	Percent of farms	Percent of litter
Methods of litter disposal:		
Applied to fi elds on the operation	60.2	39.0
Removed from the operation	70.8	60.7
Other	2.5	0.3
		100.0
Method by which litter was removed:		
Sold by the operation	33.4	36.3
Hauled off operation for a fee	4.5	4.2
Exchanged for clean-out and hauling	33.9	33.8
Exchanged for other services	5.8	5.2
Given away free of charge	21.9	20.5
		100.0

Note: Some farms have more than one method of litter disposal or litter removal, so the "percent of farms" column will not add to 100.

Source: 2006 Agricultural Resource Management Survey, version 4, production contracts only.

Farm Operators and Their Households

Farm operators make day-to-day management decisions for the farm. Until a few years ago, USDA surveys gathered information on only one operator (the primary operator) for a farm. But many farms have more than one operator, and ARMS now asks for the number of operators on a farm and a limited amount of personal information for up to three operators. The survey also gathers demographic and financial information on the primary operator's household, because important farm business and household fi nancial decisions may be linked.

We summarize operator characteristics for family owned and operated broiler farms in table 15 and compare them to operators of all family farms with at least $50,000 in gross cash farm income. Many farms, as defi ned in USDA statistics, are extremely small, with very little farming activity or income derived from farming. Almost all commercial broiler operations meet the $50,000 sales cutoff, so this cutoff was chosen as a useful comparison.

The typical operator was a white male in his 50s whose formal education ended after completing high school (table 15)—much like other farm operators. Operators of broiler farms differed from other operators in two important dimensions: more of them were women and fewer of them had graduated from college.

Eight percent of broiler primary operators were women, compared with just over 4 percent of all commercial farms, but this only captures part of the story. Most spouseoperators were women, and respondents usually report the male as the primary operator when a husband and wife each operate the farm. When we survey all operators of the farm, 54 percent of broiler farms reported having at least one female operator, compared with 36 percent of all commercial farms.

Table 15. Operator characteristics, contract broiler farms vs. all commercial farms

Item	Broiler farms		All farms, gcfi >$49,999	
	Primary operator	Operator-spouses	Primary operators	Operator-spouses
Number of farms	17,005	8,280	488,716	188,386
Operator age				
Average age (years)	55	53	55	54
Percent under age 40	8.5	10.4	11.2	10.1
Percent 65 or older	20.6	13.9	21.9	16.4
Gender and Ethnicity	**Percent**			
Female	8.0	92.2	4.4	93.0
Hispanic	0.7	0.5	1.8	1.8
African American	1.1	0.3	0.2	0.1
American Indian	1.2	0.1	0.9	0.03
Asian	1.1	1.7	0.4	0.3
Education	**Percent**			
Less than high school	13.5	9.9	7.3	8.0
High school only	49.6	52.5	38.9	35.9
Some college	24.9	24.5	25.3	29.1
College	12.0	13.0	28.4	26.8

Notes: Nonfamily farms are excluded. Table compares broiler operations with production contracts to all family farms with gross cash farm income (gcfi) of at least $50,000.

Source: 2006 Agricultural Resource Management Survey, version 4, production contracts only.

Survey respondents were asked for the highest level of formal education that they completed and given four response categories—less than high school, completed high school, completed some college, and a college degree. More than 50 percent of commercial farm operators had completed at least some college education and less than 10 percent did not complete high school. Over 60 percent of broiler farm operators had no more than a high school diploma and 12 percent completed college, compared with 28 percent of all commercial farm operators.

Farm Business Finances

ARMS collects information on farm businesses and the households of principal operators, thereby allowing users to see the links between business and household fi nances—indeed this is a primary and unique advantage of the survey. In table 16, farm-level financial data, including revenues, expense, and returns are presented for four size classes of broiler enterprises.

Organizing the Information

The data are organized into three panels. The top panel provides information on broiler enterprise characteristics for the farms in each size class. The middle panel identifi es the operation's sources of gross income and then derives net farm income from gross income and operating expenses. The bottom panel shows how net income from the farm business flows to the principal operator's household.

"Small" operations include broiler enterprises with less than 1.33 million liveweight pounds removed in 2006—the smallest 25 percent of operations. Most of those farms had 1-3 broiler houses and average production was 846,838 pounds. The "medium" class includes enterprises with between 1.33 and 3.30 million pounds removed (50 percent of all operations). Most of them had 2-5 houses. The largest 25 percent of broiler operations were divided into two equal-size classes because that group covered a wide range of enterprise sizes. The "large" class had between 3.30 and 4.49 million pounds of liveweight production, usually in 4-8 houses, while the "very large" had more than 4.49 million pounds of liveweight production. Some of the very largest operations had 18 houses, but the interdecile range (10th to 90th percentile) was 6-14.

The four classes show distinct age patterns—small enterprises have older houses (24 years, on average) and older operators (60 years), while very large operations tend to have the newest houses and the youngest operators (table 16).

Some farms, including a sixth of the very large operations, have multiple households sharing the net income from the farm, either because of partnerships or because some of the farm's equity holders are not part of the primary operator's household. We excluded those farms from the financial analyses in order to simplify the analysis of farm business fi nances and the linkage to households. Exclusion has no substantive effect on the findings presented in table 17, but it had some impact on the estimates in table 16.[15]

Net Farm Income

The middle panel of table 16 summarizes gross and net income for the farms. Gross cash farm income is the sum of fees from production contracts, revenues from cash sales and marketing contracts, government payments, and other farm income. Fees from production contracts account for most of gross cash farm income—70 percent in the small class, where some farms are diversifi ed into signifi cant production of other commodities and 80 percent or more in the other classes.

Gross farm income combines gross cash farm income with noncash income. In turn, noncash income includes the net change in inventories and accounts receivable, as well as the imputed rental value of the operator's home if it is part of the farming operation. On average, farms in each class reported substantial increases (equivalent to 5-10 percent of gross cash farm income) in inventories and accounts receivable in 2006. Industry production continues to grow, and we would expect growing industries to exhibit growing accounts receivable. Farms with crop operations could also have kept some production in storage at the end of 2006, leading to increased inventories in anticipation of rising crop prices.

Table 16. Income statement, contract broiler farms, by size of broiler enterprise

	Broiler enterprise size class[1]			
Item	**Small**	**Medium**	**Large**	**Very large**
	Farm characteristics			
Average production (lb)	846,838	2,253,019	3,845,109	6,449,921
Broiler houses (no.)[2]	1-3	2-5	4-8	6-14
Average house age (years)	24	17	14	13
Average primary operator age	60	54	52	51
	Gross and net farm income			
Fees from production contracts	44,476	112,693	191,688	322,027
plus Crop & livestock cash income	15,828	20,225	23,290	65,448
plus Government payments	1,639	2,793	2,504	5,950
plus Other farm income[3]	1,587	4,189	2,764	9,152
equals Gross cash farm income	63,530	139,900	220,246	402,577
plus Net change in inventories[4]	3,319	17,520	27,251	18,036
plus Other noncash income[5]	8,350	8,841	7,663	9,215
equals Gross farm income	75,199	166,261	255,160	429,828
minus Noncash operating expenses[6]	13,796	28,765	44,022	73,496
minus Cash operating expenses	42,681	95,236	140,276	247,884
equals Net farm income	18,722	42,260	70,862	108,448
	From farm to household income			
Gross cash farm income	63,530	139,900	220,246	402,577
minus Cash operating expenses	42,681	95,236	140,276	247,884
equals Net cash farm income	20,849	44,664	79,970	154,693
minus Depreciation expense	13,756	28,549	43,688	72,494
minus Wages paid to the operator	8	394	1,432	1,093
minus Income from land rents	135	294	339	484
equals Farm business income	6,950	15,427	34,511	80,622
times Operator HH share[7]	1.00	1.00	1.00	1.00
equals Farm business income to HH	6,950	15,427	34,511	80,622
plus HH Income from land rents	897	1,004	670	781
plus Income from another farm business	1,927	1,971	2,869	2,119
plus Farm wages paid to HH	238	1,339	2,658	2,821
equals Farm income to HH	10,012	19,741	40,708	86,343

[1]Size classes are based on liveweight pounds removed. The small class (less than 1.33 million) included the smallest 25 percent of farms, the medium class (1.33 to 3.30 million) included the next largest 50 percent of farms, and the two largest classes (3.30 to 4.486, and over 4.486 million) each had an eighth of all farms.

[2]**The row labeled "Broiler houses" reports the interdecile range**; for example, in the very large column, 10 percent of farms had 6 houses or less and 10 percent had 14 houses or more.

[3]Includes income from land rentals, custom work, machine hire, grazing, etc.

[4]Also includes changes in accounts receivable.

[5]Includes the value of farm production consumed on the farm and the imputed rental value of the operator's on-farm dwelling. [6]Includes depreciation expense charged to farm assets and noncash benefi ts for farm employees.

[7] The operator HH share is always 1.00 because nonfamily farms and farms with multiple households sharing in farm business income are excluded (8 percent of family farms have multiple households sharing farm business income).

Note: HH refers to primary operator household.

Source: 2006 Agricultural Resource Management Survey, version 4, production contracts only.

Net farm income subtracts cash and noncash operating expenses from gross farm income, and net cash farm income subtracts cash operating expenses from gross cash farm income. Cash operating expenses typically amount to 60-67 percent of gross cash farm income and average depreciation expenses, which account for almost all noncash operating expenses, amount to 30-32 percent of cash operating expenses. Net farm income is the difference between gross farm income and operating expenses, and it amounts to 25-27 percent of gross farm income in each size class.

Net farm income, however, varies widely among broiler operations, where a quarter of farms experience losses—negative net farm income. Poor productive performance may be one source of negative net income since, on average, operations with negative net farm income receive fees of 4.8 cents per pound, compared with 5.1 cent per pound for those with positive net income. Depreciation is a more important factor explaining differences in net income. On farms with negative net farm income, depreciation expenses account for 39 percent of gross income, on average, compared with 13 percent for other operations. Farms with recent major capital expenditures will usually record substantial depreciation expenses, often large enough to generate negative net farm incomes. Correspondingly, older operations with fully depreciated assets rarely report negative net incomes.

Tracking How Net Income Flows to Operator Households

The lower panel of table 16 shows how the fl ow of net income from the farm to the household is derived from the farm business's financial statement. In this case, the calculation starts with gross and net cash income (noncash income—changes in accounts receivable and inventory and the imputed rental value of operator housing—is not included). Depreciation expenses are deducted from net cash income when calculating farm income to the household.[16] Finally, note that farm income to the household also includes income from other farming activities, such as wages from working on other farms or income from another farm business. Most farm operator households do not have these sources of income, but some do.

The last row of the table reports farm income to the household. Note that it is considerably different from net cash farm income, primarily because of depreciation expenses. It also differs considerably from net farm income due to the exclusion of noncash income from farm income to the household.

FARM HOUSEHOLD INCOME

The last line of table 16 provides the farm component of household income for the primary operator's household. Note that it combines income from the farm business, the household's income from rentals of farm land, farm wages paid to household members, and income fl owing to household members from ownership interests in other farm business.

But farming is not the only source of household income, which also includes off-farm income whether earned (wages and salaries) or unearned (such as pensions or returns on financial assets). Household income is reported in table 17 for all farm households and the four size classes used earlier.

Table 17. Average household income, primary operator households, 2006

	Broiler enterprise size class				
	Small	Medium	Large	Very large	All farms
Number of farms	4,251	8,503	2,125	2,126	17,005
Average household income					
Total	$52,717	$64,974	$77,183	$130,819	$71,360
Off-farm	$42,705	$45,233	$36,475	$44,476	$43,717
Farm	$10,012	$19,741	$40,708	$86,343	$27,643

Notes: Nonfamily farms and farms with more than one household sharing in farm business income are excluded. For size class defi nitions, see footnote 1, table 16.

Source: 2006 Agricultural Resource Management Survey, version 4, production contracts only.

Several striking patterns stand out. Off-farm income is important in all size classes. On average, households earned $43,717 in off-farm income, which was substantial even among very large operations ($44,476, on average). Off-farm income accounts for nearly 80 percent of total household income, on average, for small broiler farms, but still accounts for 34 percent of the total among very large farms. We know, from table 16, that very large operations are substantial farming businesses. Gross farm income—think of it as 2006 contract fees, plus other farm revenues and increases in accounts receivable and inventories—was over $400,000, on average, at very large farms and over $250,000 among large farms. Yet, in each class, off-farm income provided a substantial fraction of total household income. Off-farm employment also frequently provides farm operator households with access to benefi ts, such as health insurance, in addition to wages and salaries.

Total household income and household income from farming rise sharply as farm size increases. While the household's income from farming averages $27,643 across all farms, it ranges from $10,012 in the small class to $86,343 among very large operations. Very large operations average nearly eight times as much production as small operations, and their nonbroiler enterprises (crop and livestock cash income in table 16) are four times larger.

Average household income for operators of broiler enterprises compares favorably to nationwide averages. Mean household income in the United States was $66,570 in 2006, compared with $72,453 for broiler farm households (table 17). Median income—half of households earn less and half earn more—was $48,201 for all U.S. households in 2006 and $56,248 among operators of broiler farms.

CONCLUSIONS

The broiler industry is organized in a unique manner. Almost all broiler production is carried out on contract grow-out operations that are linked to integrators through production contracts. Most grow-out operations are paid through a relative-performance compensation scheme in which contract fees depend, in part, on the grower's production performance relative to other growers. This organization model may be one reason for the industry's commercial success, as evidenced by high growth rates of poultry production, rapid productivity growth, and falling relative prices.

Although most contracts include a relative performance design, they are far from uniform. Contracts can include other terms that tie base payments to actions that affect grower costs or that assign some expense or revenue categories to either the grower or the integrator. Contracts also cover a wide range of specifi ed durations from just over a month to 15 years. Variations in contract design likely follow from differences in grower location, size, and housing, but substantial variation in terms and payments make it diffi cult for growers to evaluate contracts.

While the survey captures a single year of data, the bottom panel of table 16 provides a signal of changes in the industry. The smallest size class of farms have signifi cantly older houses, on average, and older operators as well. The industry is shifting away from a reliance on smaller operations and toward larger operations whose household income is more closely tied to the broiler enterprise.

The size shift is not as pronounced as the changes occurring in hog or dairy production, where very rapid shifts to much larger operations have occurred in recent years (Hoppe et al., 2007). Most broiler production still occurs on relatively small farms, as measured by gross income, and the shifts that have occurred have been more gradual.

Because newer operations are larger, they typically have a more substantial investment in housing and associated equipment. In turn, the households that own and run those operations are more reliant on the income from the broiler enterprise, and hence more sensitive to the income risks arising from energy price fluctuations and contract settlements in the broiler enterprise. New large operations typically receive longer term contracts from integrators, along with the greater assurance that longer contracts may bring, but contract compensation and reimbursement features may also need to be redesigned to handle the differing risk exposures that those larger farms face.

During the 1990s, U.S. broiler production grew at rates of about 6 percent per year, but since 2000 growth rates have decreased by half to about 3 percent per year. With slower growth, fewer new houses have been constructed and relatively few new operations have entered into production. The industry's rapid productivity growth has been driven, in large part, by improved poultry genetics, but also by steady improvements in structures, equipment, and production practices. These improvements, taken together, have led to greater feed effi ciency, lower mortality, and more intensive use of capital and labor services—essentially, raising the output of broiler meat that can be achieved for given inputs of feed, fuel, capital, and labor inputs. Slowing investment in new housing could reduce the industry's productivity growth and, in turn, limit any price advantages over other meats.

REFERENCES

Hoppe, Robert A., Penni Korb, Erik J. O'Donoghue, & David E. Banker. *Structure and Finances of U.S Farms: Family Farm Report, 2007 Edition*, Economic Information Bulletin No. 24, U.S. Department of Agriculture, Economic Research Service, June 2007.

Knoeber, Charles R. "A Real Game of Chicken: Contracts, Tournaments, and the Production of Broilers," *Journal of Law, Economics, and Organization, 5* (Fall, 1989):27 1-92.

Lasley, Floyd A., Henson, William L. & Jones, Harold B. *The U.S. Poultry Industry: Changing Economics and Structure*, Agricultural Economic Report No. 502, U.S. Department of Agriculture, Economic Research Service, February 1983.

MacDonald, James & Korb, Penni. *Agricultural Contracting Update: Contracts in 2005*, Economic Information Bulletin No. 35, U.S. Department of Agriculture, Economic Research Service, April 2008.

Ollinger Michael, MacDonald James, & Milton Madison. *Structural Change in U.S. Chicken and Turkey Slaughter*, Agricultural Economic Report No. 787, U.S. Department of Agriculture, Economic Research Service, September 2000.

Perry Janet, Banker David & Green Robert. *Broiler Farms' Organization, Management, and Performance*, Agricultural Information Bulletin No. 748, U.S. Department of Agriculture, Economic Research Service, March 1999.

Rogers, George B. "Poultry and Eggs," *Another Revolution in U.S. Farming?* Lyle B. Schertz, ed., Agricultural Economic Report No. 441, U.S. Department of Agriculture, Economic Research Service, April 1979.

Rogers, Richard T. "Broilers: Differentiating a Commodity," *Industry Studies*, 3rd Edition, Larry L. Duetsch, ed., Armonk, NY, M.E. Sharpe, 2002.

U.S. Department of Agriculture. Grain Inspection, Packers and Stockyards Administration. *Assessment of the Cattle, Hog, and Poultry Industries*, April 2005.

End Notes

[1] Other chickens include capons and stewing hens, but broilers accounted for almost 99 percent of ready-to-cook chicken production in 2006, and chicken in turn accounted for 86 percent of all poultry production, with turkey taking up most of the remainder.

[2] The report does not cover breeder broiler or pullet operations that produce eggs for hatcheries.

[3] Using Producer Price Index (PPI) data.

[4] Broiler prices rose by much less than feed prices between 1960 and 1995, as steady improvements in feed conversion offset the effects of feed price increases on costs. But broiler price changes have closely matched feed price changes since 1995, as improvements in feed conversion began to lag. In turn, retail poultry prices have risen at nearly the same rate as all food prices since 1995, after rising much less rapidly during 1960-95.

[5] We used the 2002 Census of Agriculture to identify the States: Alabama, Arkansas, California, Delaware, Georgia, Kentucky, Louisiana, Maryland, Mississippi, Missouri, North Carolina, Oklahoma, Pennsylvania, South Carolina, Tennessee, Texas, and Virginia. A focus on commercial producers of broilers raised for meat provides a large sample of similar operations for analysis.

[6] Some farms did produce birds in more than one size class during 2006 (3.2 percent of operations), but most of those were switching from one size class to another.

[7] Birds in the smallest size class (4.25 lb. or less) spend 39 days, on average, on the farm. The larger classes—4.26-6.25, 6.26-7.75, and 7.76 or more pounds—spend an average of 49, 56, and 63 days, respectively, on the farm.

[8] Newer houses can produce more birds from a given capacity than older houses because they have better temperature controls. Houses built between 2000 and 2005 produced about 8 percent more than the industry-wide average of 36.8 liveweight pounds per square foot, but the growth of new capacity still fell sharply after 2000, even when one accounts for this effect.

[9] Broiler production resembles many other economic activities in this regard: most operations are relatively small, but a signifi cant amount of production occurs on a few large operations.

[10] Contracts are, therefore, renewed frequently.

[11] Gross cash farm income includes fees from production contracts, revenues from commodity sales through spot markets or marketing contracts, government payments, and income from other farm activities such as grazing, custom work or machine hire, and land rentals.

[12] According to data from the 2004 ARMS hog version. Hog operations tend to have large crop enterprises.

[13] We also asked respondents whether they stored litter in their houses and, if so, whether they cleaned out their houses fully during 2006. About a quarter of respondents did not clean out any of their houses, although most of those crusted out their houses (removed some litter).

[14] The estimate of 4,000 birds per acre was derived from survey questions on the number of birds removed in 2006, the average weight of the birds, the share of litter applied to fi elds on the operation, and the number of acres receiving litter.

[15] Mean values of two measures (farm business income to the household and farm income to the household) would be lower if we included operations with multiple households in the calculations.

[16] This step is consistent with the practice followed in nationwide household income surveys, such as the Current Population Survey, and allows for comparisons of household incomes between farm and nonfarm households.

in: Transformation of U.S. Animal Agriculture
Editor: Justin M. Daigle
ISBN: 978-1-60876-938-4

Chapter 6

THE TRANSFORMATION OF U.S. LIVESTOCK AGRICULTURE: SCALE, EFFICIENCY AND RISKS

James M. MacDonald and William McBride
(USDA)

ABSTRACT

U.S. livestock production has shifted to much larger and more specialized farms, and the various stages of input provision, farm production, and processing are now much more tightly coordinated through formal contracts and shared ownership of assets. Important financial advantages have driven these structural changes, which in turn have boosted productivity growth in the livestock sector. But structural changes can also generate environmental and health risks for society, as industrialization concentrates animals and animal wastes in localized areas. This report relies on farm-level data to detail the nature, causes, and effects of structural changes in livestock production.

Keywords. Livestock, dairy, broilers, hogs, fed cattle, farm structure, scale economies, contract agriculture, CAFOs, growth-promoting antibiotics

ACKNOWLEDGMENTS

We thank Don Blayney and Mark Ribaudo of ERS, Jeremy Foltz of the University of Wisconsin, Rita Hansberry of USDA's Grain Inspection, Stockers and Packyards Administration, and Sherry Wise of USDA's Agricultural Marketing Service for comments on earlier drafts. Erik O'Donoghue, Richard Nehring, Carmen Sandretto, and Nigel Key, all of ERS, also contributed to the research reported on in this report. Dale Simms provided editing support.

SUMMARY

Livestock agriculture has undergone a series of striking transformations. Production is more specialized—farms usually confine and feed a single species of animal, often with feed that has been purchased rather than grown onsite, and they typically specialize in specific stages of animal production. Today's livestock farms tend to be tightly linked to other stages of production and processing through formal contracts. While the farms are usually owned and operated by a family, they rely increasingly on hired labor. And the farms that account for most production are much larger than they were in the past.

What Is the Issue?

Strong financial pressures have driven the industrialization of U.S. livestock farms. Larger operations are able to realize lower costs and higher returns, while tighter coordination among firms at different processing stages can reduce financial risks. But growing to a more efficient scale also concentrates livestock in a limited area, and excess concentrations of manure-based nutrients can lead to increased air and water pollution. Large operations are also more prone to use antibiotics intensively in order to pre-empt the spread of animal disease and to accelerate animal growth. Extensive antibiotic use in livestock raises concerns about increased pathogen resistance and related risks to human health. This report assesses the driving forces behind structural change in fed-cattle, dairy, hog, and broiler production, and describes the effects on productivity, prices, and pollution/health risks. It concentrates on changes in the size structure of farms, their organization and production practices, and their contractual links with processors and integrators.

What Did the Study Find?

U.S. livestock production is shifting to much larger enterprises, in part because of scale economies. Between 1987 and 2002, the production locus (the farm size, in annual sales, at which one half of national production comes from larger farms and half from smaller) increased by 60 percent in broiler, 100 percent in fed-cattle, 240 percent in dairy, and 2,000 percent in hog production. Recent surveys indicate that production has continued to shift to larger operations since 2002.

While most large livestock and poultry farms are family owned and operated businesses, they are becoming more closely linked to input providers and processors through formal contracts, joint ownership of animals, and vertical integration. Tighter vertical coordination can ease management of financial risks and speed the diffusion of innovations.

Structural change has led to increased productivity and, through that, to lower commodity costs of production. For example, the largest dairy farms (1,000 cows or more) had average costs of $13.59 per hundredweight in 2005, 15 percent below the average for farms in the next largest size class (500-999 head) and 35 percent below the costs for farms with 100-199 head (estimated $20.82 per cwt). Lower costs of production typically lead to lower wholesale and retail prices. However, structural change in livestock agriculture has had less felicitous effects as well.

Livestock wastes are becoming more geographically concentrated in the United States, and excessive applications of the nutrients contained in manure pose risks to air and water resources. There is a clear association between farm size and the concentration of manure—larger operations are more likely to ship manure to other operations and apply manure to their own fields more intensively. However, the cost to large farms of removing manure is still modest in relation to their production cost advantages, and there are a variety of ways to mitigate the risks from the concentration of manure. One such example is to reformulate the feed to reduce the amount of nutrients excreted by the animals.

Many hog and broiler operations provide subtherapeutic doses of antibiotics routinely in feed and water to promote animal growth and to prevent disease. The commercial value of such practices appears to be substantial in some stages of production, like nursing in hogs, but marginal in others. Other technologies, including expanded sanitation and testing procedures, can be substituted for subtherapeutic antibiotics in some stages of production.

Individual producers may have little incentive to take costly actions to mitigate the harmful effects of livestock industrialization. Livestock production is highly competitive, and operations with high costs may jeopardize their own survival in policing themselves. However, steps can be taken, at modest cost, that preserve the benefits of industrialized livestock production while limiting its harmful effects.

How Was the Study Conducted?

This report draws on recent ERS reports, which in turn rely on farm-level data drawn from the Census of Agriculture and from the annual Agricultural Resource Management Survey (ARMS). The Census provides an indepth source of information on levels and changes in farm size, specialization, and location, while ARMS surveys the financial conditions and production practices of farm businesses, and the well-being of farm households. Some ARMS versions contain detailed questions on the production and marketing practices, expenses, and revenues associated with specific commodities. This report draws on surveys of producers of hogs (in 1992, 1998, and 2004), dairy products (in 2000 and 2005), and broilers (2006).

INTRODUCTION

Livestock agriculture has undergone a striking transformation. Today, meat and dairy products typically originate on farms whose herds of cattle or hogs, or fl ocks of chickens, are much larger than in the past. These enterprises usually house a single species in buildings or in open-air pens, and provide them with feed that has been purchased rather than grown onsite. While most such farms are family owned and operated, they rely heavily on hired labor and are tightly linked to other stages of production and processing through formal contracts, alliances, and joint financing or common ownership of assets.

Most large dairy farms encompass several stages of production—in addition to producing milk, most still grow at least some of their own feed, produce and raise replacement heifers, and market dairy products and culled animals. But the beef cattle, hogs, or chickens fed on large operations are usually born elsewhere and marketed by other entities, as these grow-out farms specialize in single stages of livestock production. Producers are increasingly paid for the services that they provide, and not for the products that they sell.

This report analyzes the major shifts toward large-scale industrialized production systems in beef cattle, broilers, hogs, and milk, the four largest industries in the U.S. livestock sector. It focuses on the forces driving change and the major effects of those changes, drawing on findings from recent Economic Research Service (ERS) research that focuses on specific industries or practices. It does not assess the economics of organic or other alternative production systems, which have a growing but still small presence in the livestock sector.

Elements of Structural Change

Four elements distinguish the transformation of the livestock sector— increased farm size, changes in production technologies, increased enterprise specialization, and tighter vertical coordination between the stages of production. This report emphasizes farm size, with secondary attention to the other factors.

Most livestock are now fed in confined conditions in a barn, house, or fenced lot. Successful confinement feeding required a series of technological developments. The animals are bred to gain weight or produce milk efficiently, while also yielding specific meat or milk characteristics. Feed milling and delivery is automated, and herds and fl ocks are often grouped according to age and other characteristics and provided with feeds that are especially formulated for the group.

Another important feature of structural change is specialization. Some large farms produce only a single commodity, such as dairy farms that produce only milk or hog birthing operations that produce only nursery pigs, with no crop production. But such highly specialized operations are still the exception. Most large livestock operations also produce crops, but they increasingly specialize in a single stage of livestock production, such as hog finishing. Many of these operations may also loosen the links among their different commodity enterprises; for example, a grain and hog farm may sell its grain and purchase the feed provided to its hogs.

Today's livestock farms are increasingly reliant on contracts and other agreements to govern the links between production stages. Traditionally, farmers relied on cash markets as the primary mechanism for organizing production. They borrowed money for input expenses from lenders, raised their livestock to market weight, and sold livestock and livestock products to processors for a price determined at the time of sale.

More formal and long-term contractual relationships now cover over half of all livestock production (MacDonald and Korb, 2008). Some contracts commit processors and farmers to a specific volume of production to be delivered over time, with pricing formulas based upon product quality, volumes, and market conditions. Other arrangements pay farmers a fee for growing livestock provided by the contractor. Still others specify joint livestock ownership and financing between a farmer and contractor.

Causes and Effects of the Transformation

New technologies, which underpin modern livestock agriculture, have also driven the growth in farm size. Just as a single family can now farm far more acres than in the past, with greater yields, so too can a single family raise far more animals or poultry and realize greater yields of meat or milk.

The enabling technologies are mechanical, biological, and chemical. Larger and faster equipment allows a single producer to till, seed, fertilize, spray, or harvest more acres; to house and feed more livestock or poultry; or to milk more cows in a single day. Improvements in animal breeding have led to larger animals that gain more weight or produce more milk for a given amount of feed and labor, just as improvements in seeds have done for crop productivity. Chemical fertilizers and herbicides have increased the amount of feed that can be produced, while animal pharmaceuticals reduce mortality and increase meat or milk yields under the same level of inputs.

Many large farm businesses are run by extended families of several generations, siblings, or cousins. Others may be managed by several unrelated business partners. Larger management teams allow individual operators to specialize in crop production, herd management, or marketing, so this too can lead to increases in farm size. Improvements in information technology have facilitated the management of large-scale field operations, herd performance, and finances.

New technologies often reduce costs directly, by allowing more meat and milk to be produced for a given amount of land, feed, labor, and capital. But the new technologies also create scale economies, which reduce costs more for larger operations. As a result, larger farms realize higher profits, on average, which provides a strong incentive for operators to grow larger. In turn, lower industrywide farm costs lead to lower prices for farm commodities. Lower prices can squeeze smaller farms with higher costs, causing many to exit, to grow, or to explore niche markets for differentiated products. Lower commodity prices lead in turn to lower retail food prices, such that the benefits from technological improvements and larger farms fl ow to consumers.

While the transformation benefits society via lower food prices, it is not without costs. Large confined herds concentrate large quantities of manure, which must be removed from housing facilities, stored, and then moved to and spread on crop and pasture land. Animal manure contains nutrients like nitrogen, phosphorus, and potassium, and can therefore replace commercial fertilizers. But if not properly managed, manure can pose environmental risks. Excess nutrients do not contribute to further crop growth, but instead may damage air and water resources. Manure also contains bacterial pathogens that can pose direct threats to animal and human health.

Another environmental concern over increased scale is the widespread use of antibiotics. Large livestock operations tend to use animal antibiotics more intensively than smaller operations, as a way to control the spread of animal diseases and to promote faster growth. Antibiotics may enter natural resources through manure, and excessive use may contribute to increased resistance to antibiotics among animal and human pathogens.

Terminology: CAFOs and Large Livestock Operations

The environmental risks associated with the transformation of animal agriculture have led to ongoing discussion over the appropriate legal and regulatory responses to the risks. The farms that are the focus of this report are referred to as CAFOs (concentrated animal feeding operations) in those discussions, and the term is now in wide use.

In the U.S. Environmental Protection Agency's (EPA) designation, a CAFO is an animal feeding operation (AFO) that has been designated as a point source of pollutants. An AFO is a lot or facility where animals are confined and fed for 45 days or more in any 12-month period, and where crops, vegetation, forage growth, or post-harvest residues are not sustained in the normal growing season over any part of the lot or facility. The key elements in the definition are that the animals are confined; that they are fed, rather than grazed on grass or other vegetation; and that the "facility" refers to a structure, and not to an entire farm.

CAFOs are further defined by size. Large CAFOs are defined by animal inventories—at least 700 dairy cattle, 1,000 beef cattle, 2,500 pigs if they weigh over 55 pounds or 10,000 if they do not, and 30,000 broilers if the AFO has a liquid manure handling system or 125,000 if it does not. Medium- size CAFOs fall within intermediate size ranges and discharge wastewater or manure to surface waters, while small CAFOs are below the medium-size threshold but are designated by local permitting authorities as significant contributors of pollutants.[1]

The EPA's definition of a CAFO captures key elements of the transformations described above—a production process that concentrates large numbers of animals in relatively small and confined spaces, and that substitutes structures and equipment (for feeding, temperature controls, and manure management) for land and labor. While the EPA has a precise definition for a CAFO, the term is now used broadly and interchangeably with terms like industrialized agriculture or factory farms to refer to a production process that features confined feeding of large herds or fl ocks (Pew Commission, 2008; Union of Concerned Scientists, 2008; Starmer and Wise, 2007).

Data to Analyze the Transformation

We detail structural change with data from two large farm-level USDA databases that provide a unique and highly detailed picture of livestock agriculture in the United States. One is the census of agriculture, conducted every 5 years by USDA's National Agricultural Statistics Service (NASS). The census is an indepth and comprehensive source of information on changes in farm size, specialization, and location, and we utilize publicly available census data as well as confidential farm-level census records from the period covering 1982-2002.

The second data source is the Agricultural Resource Management Survey (ARMS), an annual survey of U.S. farms that links farm financial and production data, farm marketing and production practices, and farm household characteristics and finances. ARMS has several versions. Two focus on all farm types: a personally enumerated version (#1) that provides detailed whole-farm data and a shorter mail version (#5). Other enumerated versions target large, representative samples of producers of specific commodities; they include the whole-farm and farm household questions in version 5, but also include detailed questions on the expenses, revenues, equipment and structures, production practices, and contractual and marketing relationships associated with the commodity under study. Commodity versions are directed to producers in leading States—those that collectively account for 90 percent of production—and focus on operations whose livestock or poultry inventories exceed threshold levels (10 cows, 25 pigs, 1,000 broilers).

We use data from a 2006 ARMS broiler survey, dairy surveys from 2000 and 2005, and hog surveys from 1992, 1998, and 2004.[2] There have been no fed- cattle versions of ARMS because the concentrated nature of the industry does not lend itself to the sampling strategy used in the commodity versions.[3] As a result, our analyses of fed cattle rely solely on other USDA sources, such as census records and NASS *Cattle on Feed* reports. Because of the coverage of our data sources, we emphasize developments since 1980, with some background information on earlier developments drawn from ERS reports and other studies.

How Has the Structure of Animal Agriculture Changed?

Livestock production has shifted to much larger farms over time. We first describe those changes in size at the aggregate level, by comparing the hog, broiler, dairy, and fed-cattle industries. But each industry has also undergone important changes in how production is organized. Farms, while larger, are more likely to specialize in one stage of production and may be tied to other stages through complex contractual relationships. Size, specialization, and contractual relationships infl uence the technologies adopted by farm operators as well as the manner in which they are compensated. We describe those elements of structural change industry by industry.

Farm Size

U.S. livestock production features a highly skewed distribution of producers: many small farms raise some chickens, hogs, or cattle, but most production is on much larger farms. As such, simple measures of average size can be misleading. For example, the census of agriculture reports that 105,978 farms sold fed cattle in 2002, and that a total of 28.2 million fed cattle were sold.[4] The average farm therefore sold 266 cattle—28.2 million divided by 105,978 (table 1). But that average (the mean farm size) isn't very informative, or representative, of farms or of cattle. Most farms were much smaller than the average; 91,000 sold less than 100 cattle, and half of those sold less than 10. But most cattle came from farms that were much larger than the average: over 20 million of the cattle sold in 2002 came from farms that sold at least 5,000 head.

As production is the underlying focus of this report, we measure farm size so as to capture the farm on which the typical hog, steer, dairy cow, or broiler resides. To do so, we identify the farm size that's at the center of the distribution of production, where half of annual production comes from larger farms and half comes from smaller farms.[5] We call that operation size the locus of production.

The production locus is considerably larger than the average (mean) farm size. While the average seller of fed cattle sold 266 head in 2002, the average animal came from a farm that sold 34,494 cattle—half of all fed-cattle sales came from operations that sold 34,494 or more and half came from farms that sold less (table 1). The production locus is larger than the average farm size in all four commodities.

In each industry, the production locus increased over time, nearly doubling for fed cattle and for broilers between 1987 and 2002 (table 2). The broiler and fed-cattle industries underwent major changes before 1987 (Gee at al., 1979; Lasley, 1983; Hart, 2003), with more recent shifts of production to larger operations occurring while vertical organization remained stable.

Consolidation in the hog and dairy sectors is more pronounced. Half of all 1987 dairy production came from farms with no more than 80 milk cows in the herd. As very large operations (with 1,000 or more cows) grew more common, the locus of production grew to 275 cows by 2002. The shifts in hog production were even more dramatic. In 1987, half of all hogs marketed came from farms that sold no more than 1,200 hogs. That locus rose to 23,400 by 2002, refl ecting a major reorganization of production into stages, as well as shifts to larger operations in every stage of production.

So livestock production has shifted to much larger operations over time. Other elements of structural change—specialization, contractual relationships, technology—tend to be specific to each commodity.

Table 1. Average livestock farm size, 2002

	Mean farm size	Production locus
Fed cattle sold	266	34,494
Broilers sold	265,585	520,000
Hogs sold	2,255	23,400
Milk cows in herd	99	275

Note: The mean farm size is the total number of head sold, divided by the number of farms selling the species. The production locus measures the size of farm at which half of production came from larger farms, and half from smaller.

Source: 2002 Census of Agriculture and Hoppe et al. (2007).

Table 2. Consolidation in livestock production, 1987-2002

	1987	1992	1997	2002
	Production locus (head sold/removed)			
Broilers	300,000	384,000	480,000	520,000
Fed cattle	17,532	23,891	38,000	34,494
Hogs	1,200	1,880	11,000	23,400
	Production locus (milk cows per farm)			
Dairy	80	100 140		275

Note: The production locus measures the size of farm at which half of production came from larger farms, and half from smaller.

Source: Hoppe et al. (2007).

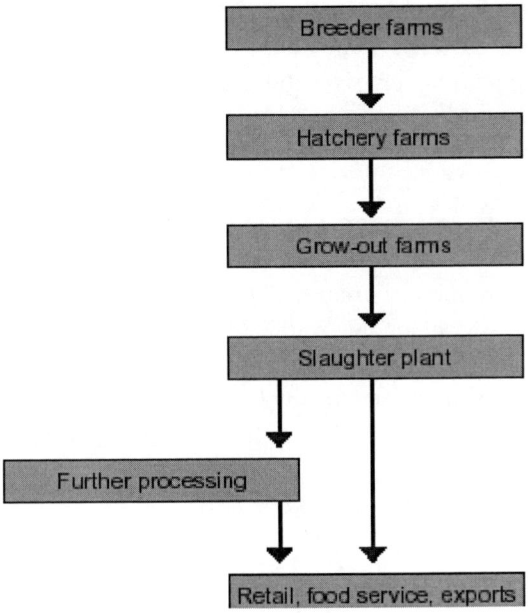

Figure 1. Organization of broiler production

Broilers

Broiler production has a distinctive organization (figure 1). Firms called integrators own hatcheries, processing plants, and feed mills. Integrators then contract with farmers to "grow out" broiler chicks to market weight, and to produce replacement breeder hens for hatcheries. Under a production contract, the integrator provides the farmer/grower with chicks, feed, and veterinary and transportation services, while the farmer provides labor, capital in the form of housing and equipment, and utilities. The birds are sent to slaughter after 5-9 weeks on the farm, and the farmer is paid for the growing services provided.

The organizational innovations developed in broiler production have been adopted in other commodities, but the methods of grower compensation remain distinctive (MacDonald, 2008). Growers receive a base payment for each fl ock of birds and an incentive. The incentive payment depends on the grower's performance, relative to other growers delivering birds to the integrator during the same period. Those growers who can convert feed to meat more efficiently, while having fewer birds die, realize higher payments. While contracts in other commodities may specify incentive payments, they are set against fixed standards and not relative performance.

The industry's current form developed during the 1950s and 1960s as integrators devised grow-out contracts, built production complexes, and developed breeding fl ocks (Hart, 2003; Lasley, 1983). Early grow-out farms weren't very large, but that changed (figure 2). In 1959, farms producing at least 100,000 broilers in a year accounted for 28.5 percent of production. That share doubled by 1969 and continued to grow rapidly until the 1990s. Today, hardly any commercial growers produce fewer than 100,000 broilers in a year. The industry's basic organization remains unchanged, but production continues to shift to larger operations, from a production locus of 300,000 broilers in 1987 to 520,000 in 2002 and 600,000 by 2006.[6]

Broiler houses are the grower's major expense. A pair of new houses can cost from $350,000 to $750,000, depending on size and location. Because of the significant initial investment in housing, lenders are important, both for funding and advice. Housing and equipment improvements are an important source of innovation and cost reduction in the industry.[7]

Measured by the gross farm income received by growers, broiler grow-out is still dominated by small farms. A farm at the 2006 production locus (600,000 broilers) would generate fees from broiler production of about $156,000 if it received the industry's average payment of 26 cents a head (MacDonald, 2008). Most realize additional farm revenues from crop production, but broilers account for most of gross income, and the production locus farm would likely realize a gross farm income, from all sources, of $180,000- $200,000. That is not a large farm—for agriculture as a whole, two-thirds of commodity sales come from farms with at least $250,000 in gross income.

Hogs

Hog production has been transformed in the last two decades. Today, operations tend to specialize in single stages of production and are linked to one another by integrators using production contracts. However, there are some important differences between broiler and hog organization.

Hog production can be described in three stages (figure 3). In the farrowing stage, sows bred by natural or artificial insemination produce nursing pigs, which are weaned at 2-3 weeks of age, when they weigh 8-12 pounds. In the nursery stage, weaned pigs are fed for about 6 weeks until they weigh 40-5 5 pounds. Finally, in the finishing stage, feeder pigs are fed for up to 6 months until they reach market weights of 250-290 pounds.

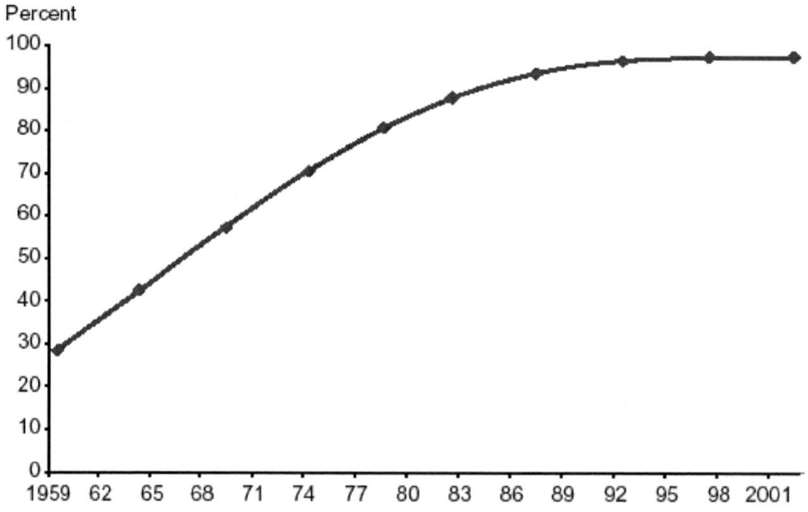

Figure 2. Share of broiler production held by farms selling at least 100,000 broilers

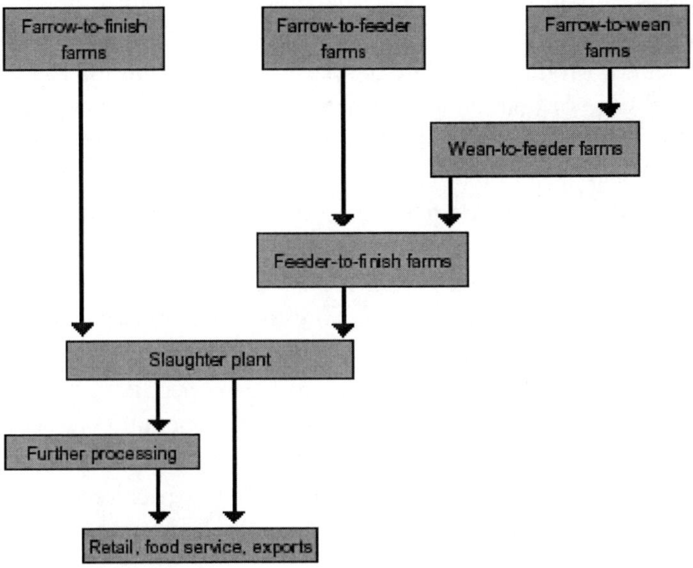

Figure 3. Organization of hog production

Table 3. Structural change in market hog production, 1992-2004

Type of operation	1992	1998	2004
Farrow to finish			
Percent of operations	54	49	31
Percent of market hogs removed	65	38	18
Percent of feed home-grown	55	51	38
Percent of hogs removed under contract	na	na	na
Percent of hogs removed under contract	na	na	na
Market hogs removed (head per farm)	886	1,239	1,472
Feeder to finish			
Percent of operations	19	31	40
Percent of market hogs removed	22	55	77
Percent of feed home-grown	45	22	15
Percent of hogs removed under contract	5	40	67
Market hogs removed (head per farm)	945	2,589	4,656

Note: Other operations, such as wean-to-finish, account for small shares of market hogs = Not available.

Source: Key and McBride, 2007.

In 1992, most market hogs still came from independent farrow-to-finish operations that combined all three stages of production and that sold their hogs to meatpackers through cash markets (table 3). Today, most production is organized by integrators who coordinate production among growers specializing in separate stages (Martinez, 2002). Some combine farrowing and nursery stages into a farrow-to-feeder operation, and some have broken the nursery stage into two. Most market hogs now come from feeder-to-finish operations that receive feeder pigs from integrators (table 3).

Some integrators organize hog production exactly as broiler integrators do—they own packing facilities and feed mills and they contract with growers to raise hogs. Others focus on organizing contract hog production, while purchasing feed from mills and selling their hogs to packers under cash sales or marketing contracts. About 40 major integrators now coordinate production of 75 percent of the 100 million hogs marketed annually in the United States, with smaller integrators and traditional producers accounting for the rest. Major integrators typically contract with providers of genetics services to help improve their breeding herds (Hart, 2003).

Farrowing and nursery operations often specialize, with no commodities other than hogs, and they can be very large. The locus of farrow-to-feeder pig production was 50,000 pigs removed in 2004, and the corresponding production locuses were 25,000 pigs removed on wean-to-feeder operations and 60,000 on farrow-to-wean operations.[8]

Hog finishing operations usually operate under a production contract with an integrator, who provides them with feeder pigs, feed, and associated services, while the grower provides housing, labor, management services, and utilities. The grower receives a fee for services, which includes a flat payment for each hog or hog space, and may also have incentive clauses tied to feed efficiency.

Finishing operations have gotten substantially larger—mean annual removals increased from 945 to 4,656 hogs between 1992 and 2004 (table 3). Most hogs come from much larger farms, and the 2004 production locus fell at 12,000 hogs. Finishing operations are usually diversified farming businesses with crop and hog production. Fees from production contracts form a smaller share of gross farm income on contract hog farms than on broiler farms (because of substantial crop production on hog farms), and the hog farms tend to have substantially larger gross farm income than broiler operations (MacDonald and Korb, 2008). While the crops grown on hog finishing operations are not generally fed to the hogs, the hogs' manure is used to fertilize the farm's cropland.

Capital requirements for a large confinement hog enterprise can be substantial. Finishing houses can cost $1 00-$ 140 per hog space, depending on local construction costs and the nature of the equipment needed. For a 1,100-head facility, that suggests an expense of $110,000-$150,000 per house. Most operations have more than a single house, and an operation at the production locus (12,000 hogs) would need 6 houses, and an investment of $600,000- $900,000. For operations in the farrowing stage, capital costs can run to $1,000 per sow (Hart, 2003).

Dairy

Vertical coordination in dairy farming hasn't changed much; most farms still market their milk through dairy cooperatives. But production is shifting to much larger farms, and there have been some important changes in how farms organize themselves.

The production locus on dairy farms rose from 80 cows in 1987 to 275 by 2002 (table 2). That comparison may even understate the nature of the shift because dairy farms cover such a wide range of sizes. Farms with at least 1,000 head in the herd accounted for 10 percent of all cows in 1992, but that share increased to 29 percent by 2002 and to 36 percent by 2007 (table 4). Most of the recent increase has occurred on farms with at least 3,000 head, according to data from the census of agriculture (MacDonald et al., 2007).

The shift toward larger farms is closely associated with some important regional shifts. Very large dairies first appeared in the West, particularly in California but also in Arizona, Idaho, New Mexico, Texas, and Washington. Milk production in those States has increased rapidly, while changing little in the traditional dairy States of the upper Midwest and Northeast, and declining elsewhere in the country. Today, very large farms account for most Western production, while production in all dairy States is also shifting toward larger herds (MacDonald et al., 2007).

Increased size also led to important changes in dairy farm organization. Whereas traditional smaller dairies relied largely on homegrown feed and pasture forage, large farms purchase a considerable amount of feed—many in Western States purchase all or most of their feed. Some large operations ship their calves to other farms, and contract with those farms to raise them to be returned as replacement heifers. And while most midsize and large dairy farms continue to be owned and operated by a family, they rely extensively on hired labor for milking crews, herdsmen, and cropworkers.

Dairy pricing, and therefore operator compensation, has always been complex, and it remains so today. Farmers are not paid for their services, as broiler and hog growers are under production contracts, but are still largely paid for their products—milk and culled cows. Milk prices can vary, even within regions, with differences in supply and demand for dairy products and because of government policy. But farmers also increasingly receive payments tied to the characteristics of the milk that they produce; most receive premiums for higher levels of butterfat, protein, and other solids, and receive less for milk with high counts of somatic cells, an indicator of bacteria.

Beef

Producers specialize in distinct stages within the cattle industry, and those stages are organized and coordinated differently (figure 4). The cow-calf stage of beef production includes many farms: 758,000 operations had beef cows in 2007, 10 times as many as had dairy cows. Moreover, there are few really large cow-calf operations. Western cow-calf operations tend to specialize in cattle raising, while those in the Midwest and South tend to combine a cattle business with a crop enterprise. While production has shifted to larger operations over time, cow-calf production is still spread over many modestly sized farms and ranches.

Most calves are weaned between 6-9 months of age, when they weigh 400-700 pounds. They may remain on pasture at the farm or ranch, or be sent to pasture elsewhere. Most move to specialized stocker or backgrounding operations that add 200-400 pounds of weight over 3-8 months. Such operations provide a marketing function by assembling groups of animals of consistent quality, and they can also improve quality through health and nutritional management of the cattle, often with a combination of pasture forage and feed.[9] Most cattle will eventually be moved to feedlots, where they are confined outdoors in pens with other feeder cattle and fed a highenergy ration (RTI, 2005). Feedlots are the industrialized stage of the sector.

Table 4. Size structure of dairy farms, 1992-2007

	1992	2000	2007
Herd size (milk cows)	Percent of cow inventory		
1-49	20.4	12.0	7.4
50-99	29.0	22.0	15.4
100-1 99	19.0	18.0	13.4
200-499	13.7	16.7	14.9
500-999	8.0	12.0	12.5
1,000+	9.9	19.3	36.4

Note: Herd size refers to all dairy cows on an enterprise, including dry cows but excluding calves, heifers, and bulls.

Source: USDA/NASS *Milk Production*, February Issue (1992 and 2000); USDA/NASS *Farms, Land in Farms and Livestock Operations* (2007).

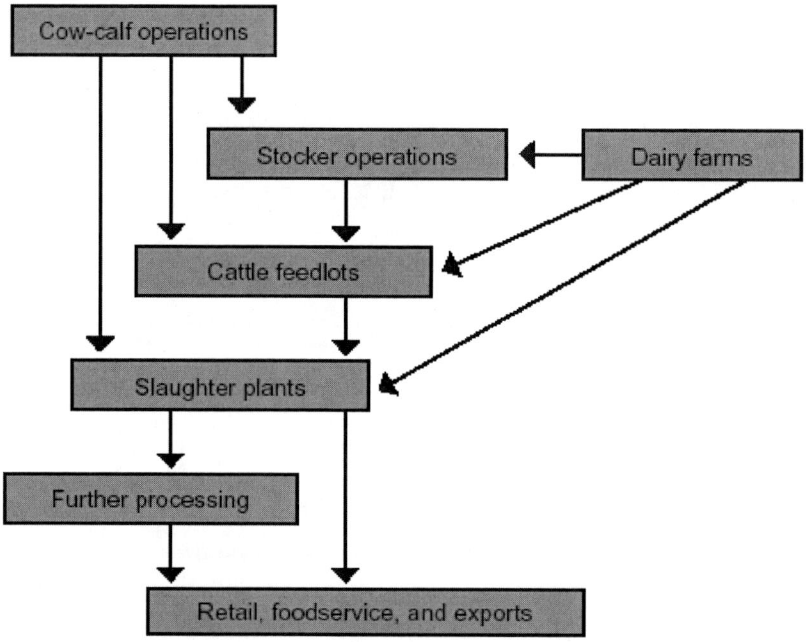

Figure 4. Organization of beef production

Until the mid-1960s, most feeding occurred in small "farmer-feedlots." These farmers, usually in the Midwest, raised their own feed for cattle that they purchased and fed to market weight after harvest, when they had spare time. In 1964, feedlots with capacities of less than 1,000 head handled over 60 percent of U.S. fed-cattle marketings.

Cattle feeding underwent major changes in the 1960s and 1970s. Production shifted from farmer-feedlots to commercial cattle feeding businesses in the Great Plains and the West (figure 5). Large commercial feedlots purchase all or most feed ingredients and maintain feedmills, feed delivery systems, and information systems onsite. They employ nutritionists, veterinarians, and sales and management staff, and they buy and sell cattle weekly.

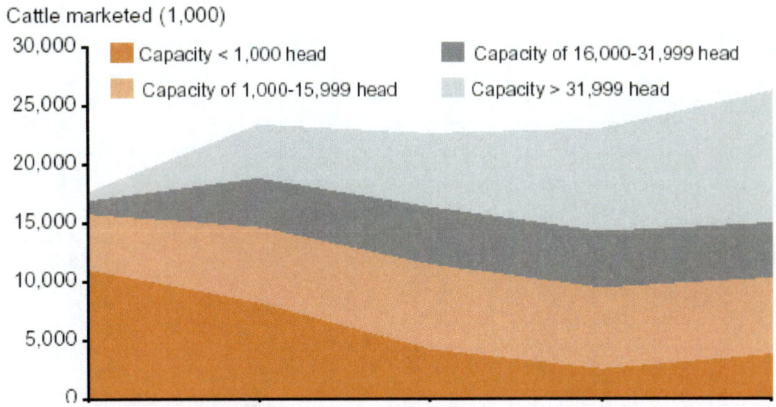

Figure 5. Fed cattle shift to large feedlots

While there are still many farmer-feedlots across the country, such operations feed only one-third as many cattle as they did in the 1960s. Meanwhile, commercial feedlot production has concentrated in larger businesses. In 2007, 262 feedlots had a capacity of at least 16,000 head, and they handled 60 percent of U.S. fed-cattle marketings. The largest can feed 100,000 cattle at a time. Some are owned by meatpackers, some are part of larger diversified firms, and others are specialized cattle feeding businesses, sometimes with a feed production enterprise as well.

Cattle feeders own cattle and market them under their own account, but they also **"custom-feed" cattle for others through production** contracts. Under those contracts, feeders are paid for their services and an owner may market the cattle or have the feeder handle that service as well. Contractual relationships are becoming more complicated as backgrounders or cow-calf operations enter into joint ownership of cattle with feedlots or with processors. Joint ownership provides more sources of financing, and also ties compensation more closely to cattle quality at each stage of the production process.

DRIVERS OF STRUCTURAL CHANGE: TECHNOLOGY AND SCALE ECONOMIES

Production has shifted to much larger farms in the broiler, dairy, fed-cattle, and hog industries. To what extent do costs account for those shifts? If large farms do realize cost advantages, do they follow from technological scale economies? How large are the scale effects, and is there a level beyond which further size increases have little impact on costs? In evaluating the evidence for each commodity, we find that scale-related cost advantages are important factors in structural change, although the magnitude of the advantages, and the strength of the evidence, varies across commodities.

Multiple ARMS hog and dairy surveys provide data to support detailed cost analyses, for periods in which each sector was undergoing important changes. Our evidence for fed cattle and broilers is more limited. We have a single ARMS broiler survey and no fed-cattle survey, and major structural change occurred in the more distant past for those two commodities.

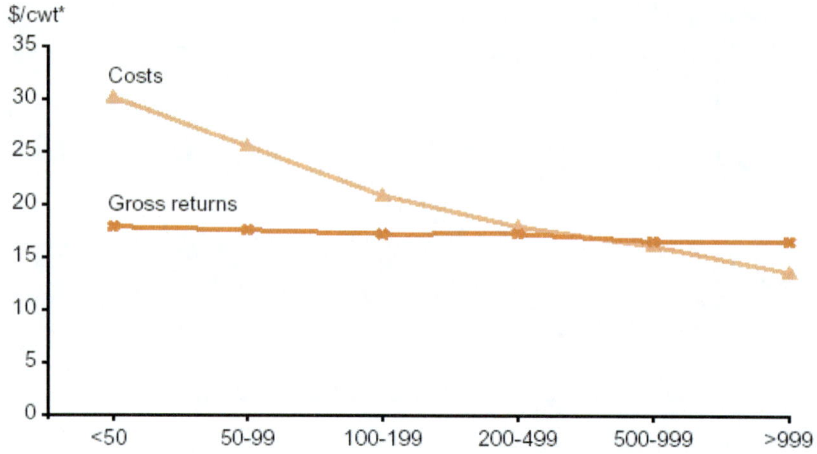

Figure 6. Average costs and gross returns in 2005, by size of dairy herd

Dairy

ERS estimates of dairy costs of production show that larger farms had substantial cost advantages, on average, over smaller operations (see box "Measuring Hog and Dairy Costs and Returns" for details on how the estimates are constructed) in 2005. Farms in the largest size class—herds of 1,000 or more milk cows—had average costs of $13.59 per hundredweight in 2005 (figure 6), about 15 percent below the average for farms in the next largest size class (500-999 head) and 35 percent below the estimate ($20.82 per cwt) for operations with 100-199 head. Average costs are much higher among even smaller operations.[10]

One source of scale economies in dairy is capital equipment—large and highly automated milking parlors and feed delivery systems. Structures are a second: modern free-stall barns allow large operators to realize lower housing costs per cow. Large, automated manure removal and storage systems also appear to reduce costs of manure handling, per ton, compared to smaller and less automated systems. These facilities allow larger operations to greatly economize on labor and capital costs, per pound of milk produced, by intensively utilizing their equipment and structures. On many small dairy farms, facilities are utilized at less than full capacity, especially if the facilities are nearing the end of their useful life or if the operator is nearing retirement, with no succession plans.

Small farms tend to be located in regions with higher milk prices, so they realize slightly higher average gross returns to milk production than large farms (figure 6).[11] However, costs still exceeded gross returns, on average, for small farms in 2005, while farms with at least 500 head had gross returns that exceeded their costs.

ERS cost-of-production accounts allow us to assess several different measures of the financial returns to dairy production, and to compare them across farm sizes (figure 7). Operating costs include expenses for feed, bedding and litter, fuels and electricity, and veterinary services. While farms may fail to realize enough revenue to cover operating expenses over short periods (because of unforeseen emergencies), they won't stay in business for long unless they cover such expenses. No more than 15 percent of farms in each size class failed to cover operating expenses (figure 7).

Measuring Hog and Dairy Costs and Returns

ERS estimates include all costs and returns associated with production, including those borne by integrators and landlords as well as those borne by farm operators. Gross returns include cash receipts received from the sale of hogs, pigs, or milk; the value of livestock removed under production contracts; and the value of secondary products such as culled animals and manure. Cash receipts are reported directly in ARMS. Commodities removed under a production contract are valued using production reported in ARMS and State-average prices. Actual cash receipts are reported for some secondary products, while others are valued with reported production and State-average prices.

Some components of commodity production costs—like purchased feed, feeder animals, hired labor, bedding and litter, and fuels and electricity—are reported directly in the surveys. But significant implicit expenses are also incurred on farms. For example, farm operators and their families contribute their labor to the enterprise, but since that labor often does not receive a wage, an explicit expense can't be recorded. Nevertheless, the opportunity cost of the labor should be recognized since those hours could have earned income in another activity, such as working off the farm.

ERS estimates an opportunity cost of unpaid labor, based on the off-farm labor earnings of farm operators as recorded in version 1 of ARMS, which provides data on the annual hours worked and wages and salaries earned off-farm by respondents.

Hog and dairy farms can also incur implicit expenses for homegrown feed and for capital equipment and structures. ERS uses market price data from other USDA sources to value the quantities of homegrown feed and forages produced on the farm and fed to animals, as reported in ARMS. ERS also estimates the annualized cost of replacing the capital used for livestock housing, feed and manure storage structures and handling equipment, milking facilities, tractors, and trucks, as well as the return that the capital could have earned in an alternative use. Survey respondents report the type, capacity, and characteristics of different types of equipment and structures in the enterprise. ERS analysts add secondary information on acquisition prices and useful lives of various types of capital, and interest rates, to estimate annual capital replacement costs.

ERS reports gross returns and total costs, per cwt of milk for dairy and per cwt of liveweight gain for hogs. Total costs are sorted into operating and overhead costs. Operating costs are further sorted into feed and other costs, while allocated overhead includes labor, capital recovery, the implicit rental rate of land, taxes and insurance, and general farm overhead. Further details, and data, are available at www.ers.usda.gov/Data/CostsAndReturns/

A second standard is to compare gross returns to all costs except those for capital recovery. Farms that meet that standard realize enough revenue to cover all operating expenses, all taxes and insurance on the operation, and the opportunity cost of the operator's time. Existing dairy farms have already put their capital structure and equipment in place; if they are not earning enough to pay for the replacement of that equipment, they can still continue operating until the equipment wears out, which can be a long time.[12] Only about 20 percent of farms with fewer than 100 head meet this standard, while 50-70 percent of midsize farms (100-499 head) and 80-90 percent of large farms do (figure 7). The smallest dairy operations face seemingly strong pressures to exit, as an operator could earn higher returns to his or her labor elsewhere.

Finally, total costs include capital recovery costs. Farms that cover total costs earn more for their operators' labor than the operators can earn elsewhere, and return more on the farm's capital investment than could be earned elsewhere. Over 70 percent of the largest dairies had gross returns that exceeded total costs, compared to less than 10 percent of small farms and less than 40 percent of midsize farms (figure 7). That pattern of performance provided strong incentives for existing dairies to expand or to exit, and for producers entering the business to enter at a large size.

Hogs

Key and McBride (2007) and McBride and Key (2003) provide a comprehensive overview of hog production costs and productivity. In 1998, large hog operations had substantial productivity advantages over smaller operations—they used much less feed, labor, and capital for every hundred pounds (cwt, again) of hog production.[13] Many of the highest cost operations closed after 1998, and more efficient small operations were still at a productivity disadvantage in 2004.

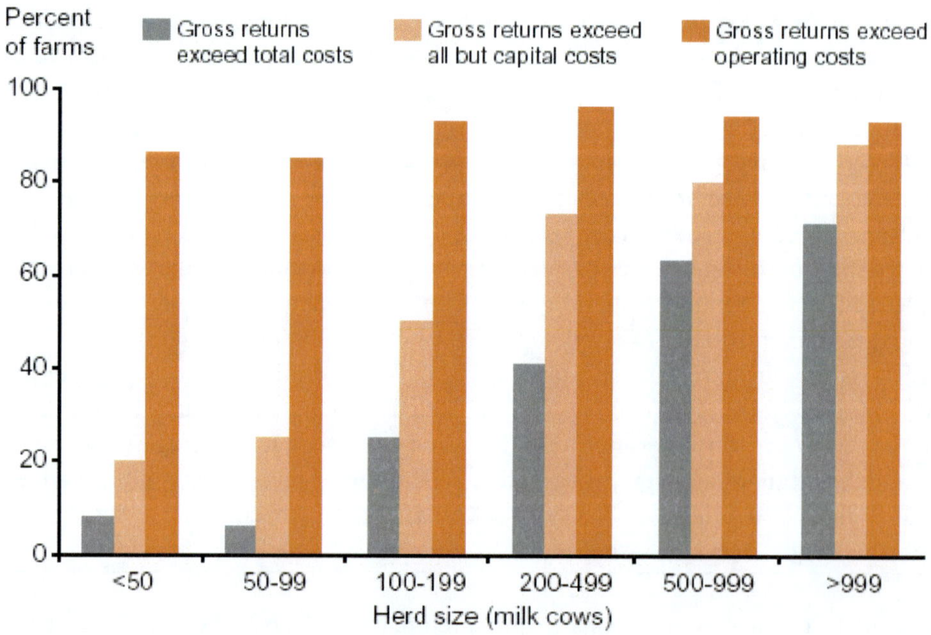

Source: 2005 Agricultural Resource Management Survey, version 4.

Figure 7. Financial performance, by size of dairy farm

Productivity differences translate into important cost advantages for larger operations. Production costs, for farrow-to-finish and feed-to-finish operations of different sizes in 2004, are compared in figure 8. Costs are expressed in dollars per cwt of weight gain: for example, a finishing operation that receives 50-pound pigs and feeds them to 250 pounds would realize 2 cwt (200 pounds) of gain for each market hog removed. ERS cost-of-production estimates include all costs incurred, whether borne by the integrator or the grower. Feeder pig costs are excluded from the estimates of costs at feederto-finish operations, so one should not compare costs across types; instead, the proper comparisons are among different size classes of each type.

Among farrow-to-finish operations, costs fell quite sharply, from over $70 per cwt to just over $40, as annual output expanded from 100,000 pounds of liveweight production to 1 million pounds (or from 400 hogs to 4,000, for 250 pound hogs). Costs continued to fall after that threshold, but more modestly, to $36 among operations with at least 2.5 million pounds of live- weight production (10,000 hogs).

Feeder-to-finish operations also show a strong association between costs of production and the operation's size. Average production costs fell from $45 per cwt, at 100,000 pounds of production, to $23 per cwt at 2.5 million pounds. Beyond that threshold, costs vary little (figure 8).[14] While there is a range of actual costs around the averages, the evidence from the hog surveys indicates that large industrialized hog operations hold substantial cost advantages over smaller farms. The scale effects may be even stronger in more complex statistical analyses that control for location, production practices, and operator characteristics (McBride and Key, 2003: McBride, Key, and Mathews, 2008).

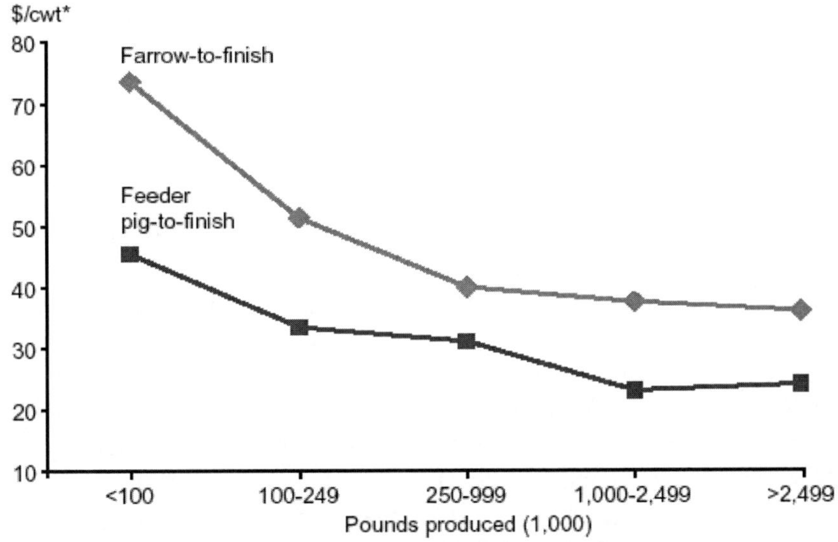

* Production cost per hundred pounds of weight gained while on the farm.
Source: Key and McBride, 2007

Figure 8. Hog production costs in 2004, by size and type of operation

Broilers

Broiler production requires a significant investment in physical capital. Farms use specialized broiler houses, which include automated equipment for providing feed and water to the birds as well as sophisticated climate control systems. Modern farms also use mechanized equipment to gather broilers for shipment to processing plants and to remove litter from the houses. Scale economies arise from innovations in the design and utilization of structures and equipment, which also allow for more effective use of labor. While scale economies in poultry processing are large and extensive (Ollinger et al., 2000), those in broiler grow-out are modest, but they have increased over time.

The most common broiler house built in the last decade covers 20,000 square feet (40 feet wide and 500 feet long).[15] Given typical flock turnover, a single house of that size can produce 115,000-135,000 birds in a year, depending on bird size. However, further economies in the use of feeding and watering equipment, as well as litter removal and storage, can be realized by operating houses in pairs, which raises the minimum efficient scale of an operation to 230,000-270,000 birds a year. Few houses built in recent years are less than 20,000 square feet, and an operation with production well below 230,000 head would see noticeably higher costs arising from limited capacity utilization, or from the higher per-unit costs of building smaller facilities.

Some grow-out operations have up to 18 houses, but the major technology- related scale economies lie in house size and feeding equipment, so most of the available advantages of scale can be realized at what is a modestly sized operation today. Almost all production in the 1950s and 1960s came from much smaller operations than today. The development of new scale economies in grow-out houses and equipment has driven the gradual increases in broiler farm size.

Beef Cattle

Cattle feeding requires a significant investment in physical capital. Feedlots use mechanized equipment for feed milling, handling, and storage, and for manure removal, storage, treatment, and transport. They use vehicles for transporting animals, pens for holding them, and other specialized structures (Martin, 1979). Equipment and structures are subject to economies of scale from two sources. First, although larger facilities cost more than smaller facilities, their cost per unit of capacity is often lower. Second, once they are built, the fixed costs of the capital can be spread across more animals if the lot can operate near capacity, so that automation favors year-round intensive use of facilities. As a result, commercial feedlots became year-round operations, while farmer-feedlots often operated seasonally.

Krause (1991) reviewed studies of feedlot scale economies that were performed during the industry's transition in the 1960s and 1970s from small farmer-feedlots to large commercial operations. Two findings stand out, and are common to other studies of livestock feeding costs. First, substantial scale economies, up to a threshold size level, derive from more intensive use of equipment, structures, and labor. Second, above the threshold there were a wide range of large sizes over which costs varied modestly, and did not rise. This bears out the wide range of commercial feedlot sizes that we see persisting to this day.

Our best evidence and most complete data—that for hogs and dairy—shows that costs fall sharply as enterprise size increases, up to some threshold level. Technological scale economies, associated with structures and equipment, seem to account for the cost advantages. The broiler and fed-cattle industries also appear to be subject to size-related cost advantages over some range of production. Moreover, none of the livestock industries appear to display a unique optimal size, once scale economies are exhausted. Farms with 2,000 dairy cows have average costs similar to farms with 1,000 cows, as do farms with 12,000 or 5,000 finishing hogs. Farms can get quite large, without real izing scale diseconomies.

Other Drivers of Structural Change

This report has focused on technological economies of scale as a major factor in structural change, and has therefore presented data on the links between the size of an operation and costs. But other factors may also matter. We first focus on complementarities among stages—the possibility that large processing plants may give rise to large livestock farms, independent of any scale economies. We next discuss the argument that Federal commodity payments for feed grains implicitly subsidize large livestock operations, thereby causing structural change. Finally, many small farms continue to survive and prosper despite the cost advantages accruing to large farms, and we evaluate some factors behind their survival.

Com Plementarities among Stages

The shift to larger livestock farms has occurred at the same time that livestock and poultry slaughter plants became much larger so as to realize scale economies and lower processing costs (MacDonald et al., 2000; Ollinger et al., 2000). Larger, more automated processing plants must obtain large and steady fl ows of uniform animal and bird types if they are to realize any potential scale economies (RTI, 2007a). Different strategies have been devised to manage those flows, but they all rely on tighter coordination of the production process. They may also encourage larger farms.

Hog and broiler integrators achieved steady flows of uniform animals to their plants by directly controlling the production process. They time chick and pig placement on grow-out farms so as to optimize flows to processing plants 6 weeks (small broilers) to 6 months (hogs) later. They can also realize uniformity by controlling the genetics of their pigs and chicks and by controlling the length of the feeding period.

Large cattle slaughter plants and large feedlots emerged during the 1970s and 1980s in the same geographic areas of the Great Plains. Beefpackers do not use the tight system of integration and production contracts that broiler and hog firms use. Instead packers use a combination of packer ownership and financing, long-term marketing agreements, and cash market purchases to manage cattle flows to plants.[16]

The coincident timing of structural changes to large processing plants and large production facilities suggests a complementary relationship between them. Each needed an opposite party willing to provide or receive large flows of uniform animals, shipped on a regular basis, so that they could reduce the risks of large-scale investment. And each needed some mechanism, such as contracts or other long-term commitments, to make the process work. In that sense, the emergence of large, capital-intensive processing plants may have encouraged a shift to larger and more specialized farms, aside from any internal scale economies in those operations.

Input Prices

Large and small farms use different combinations of inputs. For example, large dairy farms use more purchased feed and less homegrown feed and pasture forage than small farms. These differences imply that input price changes can affect farms differently. For example, increases in purchased feed prices or wage rates for hired labor would raise costs more at large farms than at small farms, because those inputs typically account for higher shares of large farm costs.

Two recent reports argue that Federal commodity programs reduce prices for a particular input, purchased feeds, and that those reductions drive structural change (Pew Commission on Industrial Farm Animal Production, 2008; Union of Concerned Scientists, 2008). Specifically, they assert, drawing on Starmer and Wise (2007), that policy encourages increased production of feedgrains, that the increased production substantially reduces feedgrain prices, and that the buyers of purchased feed benefit from lower feed prices. While producers of homegrown feed have been the direct recipients of commodity payments, the reports argue that the payments have not been large enough to offset lower commodity prices, so that commodity programs have largely benefited large-scale animal feeding operations at the expense of smaller diversified crop and livestock farms.

However, the size-related differences in production costs that are summarized in figures 6-8 cannot be attributed to the effects of commodity programs. In developing cost-of-production estimates, ERS prices homegrown feed at its market value—that is, the price that homegrown feed would have drawn as feed in regional feed markets—and does not attempt to estimate the actual cost of producing the feed. To the extent that larger operations realize lower feed costs in ERS estimates, it is because they use less feed per cwt of production and not because purchased feed costs them less.

But even if commodity policies reduce purchased feed prices, there's no reason why that should alter the size structure of livestock farms. No technological barrier prevents small farms from replacing homegrown feed with purchased feed: if purchased feed prices are lower than the costs of growing feed, then small livestock operations can simply buy feed and realize the same savings as large farms.[17] A difference between the price of purchased feed and the cost of homegrown feed does not explain why feeders build large rather than small operations.

Many Small Operations Survive

Even as cattle feeding has shifted to large commercial feedlots, small feedlots of less than 1,000 head capacity still feed nearly 4 million cattle a year, and their share of the fed-cattle market has stabilized or even grown lately (figure 5). Similarly, while large dairy farms hold substantial cost advantages over smaller farms, and production continues to shift to larger operations, many small farms remain in business and some are profitable (figure 7).

Some of the size-specific variation in farm financial performance refl ects unexpected events--related to weather, disease, accidents, or market changes—that temporarily affect revenues or costs. But other factors may be more systematic. Farms may persist in spite of poor financial performance because operators are willing to accept lower earnings than they could earn elsewhere. Others may have underutilized assets, in capital or the farm operator's time, that would not otherwise be used except in the livestock operation (this is a traditional reason for feeding small lots of cattle outside of busy growing seasons). In this case, ERS estimates may overstate the true opportunity cost to the enterprise of using the inputs.

Some farms can realize substantially higher revenues than others because they are in a favorable location, or because they produce a niche product. For example, large hog operations usually produce hogs bred and fed to gain weight quickly and efficiently. The low-cost pork derived from those animals may not have the flavor or texture that some buyers seek. Smaller operations that specialize in specific breeds may have higher production costs, but can still prosper if they find enough buyers willing to pay a premium for a differentiated pork product.

Substantial differences in managerial and technical skills may also affect farm financial performance. Better farmers likely maintain lower costs and perform better financially. Some small operations remain profitable simply because their operators are unusually good at their jobs.[18]

IMPACTS OF STRUCTURAL CHANGE

Livestock production is shifting to much larger farms. Farms are specializing in stages of livestock production, and contracts are increasingly being used to coordinate the actions of farmers in different stages. These shifts, often quite dramatic, have important impacts. To the extent that scale economies and greater efficiency are driving the shifts to large size and to contract production, then the shifts will lead to improved agricultural productivity— more meat and dairy products for a given commitment of land, labor, and capital resources—and lower wholesale and retail prices for meat and dairy products.

But structural change leads to other impacts as well. Structural changes in farming and processing, toward larger plants and fewer buyers, have occurred in tandem. Larger farms concentrate animals in small areas, heightening pollution risks from excess manure nutrients in land, water, and air resources. Finally, farmers with large herds or fl ocks in confined areas are more vulnerable to the rapid spread of animal diseases, which they combat with the widespread use of animal antibiotics. This has led to concerns that such widespread use creates human health risks if animal antibiotics accelerate the development of resistance among human pathogens.

Productivity and Prices

Major structural changes often generate large increases in productivity. Increased productivity reduces production costs, leading to lower commodity and retail prices. Productivity improvements also free land, labor, and capital resources that can then be used to produce more of the commodity or of other goods and services.

The broiler industry improved its productivity greatly as its vertically integrated system of production was introduced in the 1950s and then refined through improvements in breeding, feed formulations, housing, and management practices. In 1955, when modern integrated broiler complexes were being introduced (Hart, 2003), it took 73 days to produce the average broiler, which weighed 3.1 pounds, and every 100 pounds of broiler production required 285 pounds of feed and 4 hours of labor. By 1980, it took 52 days to produce a broiler that weighed about 4 pounds, and every 100 pounds of broilers required 208 pounds of feed and 30 minutes of labor (Lasley, 1983).

Because broilers could be produced with less feed, labor, and capital (less time means less capital per broiler), broiler production costs were only slightly higher in 1980 than in 1955, even though prices for production inputs had risen. As a result, retail prices for broilers rose by only 30 percent between 1955 and 1980, while the overall consumer price index for food more than tripled.[19]

Improvement continued after 1980, but at a slower rate. By 2006, it took 49 days to raise the average bird, which weighed 5.4 pounds, and producers used 195 pounds of feed for every 100 pounds of broiler production.[20] With slowing productivity growth, retail chicken prices increased only slightly less than beef or food prices after 1980.[21]

We have more limited data on cattle productivity, and existing studies use annual industrywide data to analyze the drivers of change in prices paid for feeder cattle. Marsh (2001) measured beef cow productivity over time as annual beef production divided by the Nation's inventory of beef cows (adjusted for imports). Beef production per cow can be increased by raising birth rates of live healthy calves, reducing cow and calf mortality, increasing the mature size of fed cattle, and speeding the time it takes to reach market weight. While the steps to increase birth rates and reduce cow mortality occur in the cow-calf stage, the other routes to boosting productivity may occur in the fed-cattle stage.

Marsh's measure of beef cow productivity increased by 35.5 percent, or 1.6 percent per year, from 1980 to 1999. He attributed the growth to improvements in breeding, animal nutrition, and animal health in both cow-calf operations and feedlots. Productivity growth led to reduced costs and lower feeder cattle prices over the period.

Hog sector structural changes occurred more recently. Key and McBride (2007) compared industry-average production costs for hogs in 1992, 1998, and 2004, with separate estimates for feeder-to-finish and farrow-to-finish operations. As with broilers, industrywide efficiency improvements reduced the amount of inputs needed to produce pork. In the feeder-to-finish sector, average feed conversion rates (pounds of feed required for every cwt of weight gain) fell from 383 pounds in 1992 to 214 pounds in 2004. Labor use (per cwt of weight gain) fell from 53 minutes to 9 minutes. As a result, average production costs fell from \$37.54/cwt in 1992 to \$26.59/cwt in 2004, despite increases in the prices paid for capital, labor, feed, and other inputs. In real terms (all input prices in 2004 dollars), industrywide average costs dropped by 44 percent between 1992 and 2004.

Key et al. (2008b) estimated that productivity on feeder-to-finish farms grew at an average annual rate of 6.3 percent from 1992 to 2004. When compounded over 12 years, that estimate implies that the amount of hog production that could be realized from a given set of feed, labor, capital, and other inputs more than doubled. Key et al. attributed nearly half of the growth to the realization of scale economies by larger farms and almost all of the remainder to innovations in breeding, feed formulations, and equipment that raised productivity across all farm sizes.

Milk production has benefited from ongoing improvements in breeding, feed formulations, equipment, and management practices, all of which are independent of shifts to larger farms. Still, MacDonald et al. (2007) estimated that industrywide milk production costs were 8 percent lower in 2006 than they would have been had production not shifted to larger and lower cost farms during 2000-2006. The analysis held input prices constant, so the estimated cost declines did not refl ect lower prices for inputs, but instead reflected less labor and capital used for a given amount of milk production on larger farms. The calculation captures a short window of time, and the effects over a 10- to 20-year window would likely be commensurately larger.

Structural changes in each livestock commodity have led to substantial cumulative reductions in the resources needed to produce a given amount of meat or milk, and to lower producer and consumer prices. But other consequences have been less felicitous.

Processor Concentration

As farm production shifted to larger operations and tighter vertical linkages, the livestock processing industries became much more concentrated. Most producers now face just a few buyers for their livestock, livestock products, or grower services. While mergers among processors have played a role in increased concentration, the major factor has been increases in plant size which have allowed processing plants to realize scale economies and lower costs (MacDonald et al., 2000; Ollinger et al., 2000). If those lower costs reduced retail prices and led to increased quantities demanded, growers could benefit from increased processor concentration. But increased concentration could also confer market power on processors, enabling them to impose lower prices on growers and higher prices on consumers.

Table 5 summarizes concentration trends among cattle, hog, and broiler processors.[22] The four largest meatpackers held 79 percent of steer and heifer slaughter in 2005, up from 36 percent in 1980, an extraordinary increase that attracted widespread attention.[23] Concentration ratios in cow and bull, hog, and broiler processing have also risen steadily and are substantially higher than in 1985.

National concentration measures understate the concentration that many farmers face in local and regional markets. For example, although feeder animals may move across North America, fed cattle and market hogs rarely move far to slaughter, and feedlots and market hog sellers usually have access to just three or four buyers.

Larger plants allowed packers to realize lower processing costs through economies of scale, with the gains from lower costs passed to consumers and to farm operators (Brester and Marsh, 2001; Ollinger et al., 2000; MacDonald and Ollinger, 2005). However, production is no longer shifting to large processing plants, and there are widespread concerns that packers and integrators may be able to exploit high levels of concentration and reduce the prices that they pay for livestock, livestock products, and grower services.

Highly concentrated local markets for agricultural products aren't new, and some widely used agricultural marketing institutions are designed to limit the exercise of market power and to induce farmers to commit to production in markets with few buyers. For example, farmer-owned cooperatives play a major role in milk handling and processing—70 percent of producers, accounting for 76 percent of milk production, ship their milk to plants operated by their own cooperative.[24]

Many fed cattle are sold to packers in cash markets where prices are determined on a daily and weekly basis. Most smaller producers, and most small packers, use cash markets exclusively, while large producers and packers use a variety of marketing channels (RTI, 2007a). The market is characterized by high buyer concentration, and by several features of bidding that would seem to discourage aggressive competition among packers (MacDonald, 2006; Xia and Sexton, 2004).

Because of its size, high concentration, and readily available data, there are many studies of competition and pricing in the fed-cattle industry. Some find that prices look like those that would prevail in competitive markets—that is, they find no evidence of packer market power (Morrison-Paul, 2001). Others find that prices are below those that would prevail in competitive markets— that is, they find evidence of monopsony on the part of packers—but the estimated effects on prices are small (Azzam, 1997). Despite high concentration and bidding practices that might facilitate the exercise of market power, the studies find little evidence of extensive market power in fed-cattle (Koontz, 2003). However, increases in concentration from current levels could provide packers with the power to force cattle prices down. In markets with only three or four buyers, mergers among buyers will attract close scrutiny, particularly where there are no apparent offsetting efficiency gains.[25]

Growers of broilers, and growers of hogs under production contracts, operate in a different environment. They do not sell livestock in commodity markets, but instead are hired to grow for an integrator who links stages in a production complex that covers just a few counties. The competitive issues in these businesses, therefore, refer to labor markets for grower services and not to product markets for livestock. While hog growers tend to have three or more integrators operating in their area, sometimes with cash market alternatives, broiler growers are likely to have only one or two integrators, with no cash market outlets (MacDonald and Korb, 2008).

Table 5. Concentration in poultry and livestock procurement

Share of purchases by four largest processors						
Commodity	**1980**	**1985**	**1990**	**1995**	**2000**	**2005**
Cattle	Percent					
Steers and heifers	36	50	72	79	82	79
Cows and bulls	10	17	20	23	32	49
Hogs	34	32	40	46	57	64
Broilers	na	34	41	46	49	53

Source: USDA Grain Inspection, Packers and Stockyards Administration (GIPSA), *Assessment of the Livestock and Poultry Industries* (various years).

In markets for grower services, potential hog or broiler growers have choices—they can invest their time or money in other agricultural commodities, or in nonfarm employment—and integrators must compete with those activities to attract growers. Production contracts commit the integrator to provide chicks or pigs, and commit the integrator to a compensation agreement--contracts therefore offer more assurance to growers than they would get in a highly concentrated cash market, where no such assurances exist.

The competitive risk to contract growers arises after they have made a substantial capital investment in housing, when the initial contract expires. Many production contracts are short term, as little as the 3 5-45 days it takes to grow a fl ock, even though the grower makes a long-term financial commitment. At that point, an integrator may be able to impose extra costs or lower payments on a grower as a condition of contract renewal. However, an integrator that develops a reputation for exploiting growers will have difficulty attracting new growers. And growers—especially hog producers who have more options—may contract with other integrators or shift to cash markets.

Spot Markets and Marketing Information

Spot market transactions provide benefits to all market users because transaction terms—including prices, locations, and observable quality characteristics—are readily available and easy to record. Accurate and widely available market information helps speed the discovery of a "market price" through arbitrage, as buyers try to avoid paying unnecessarily high prices and sellers try to avoid accepting unnecessarily low prices. Visible, accurate, and reliable market price information should provide important signals such as value differences, regional price differences, and quantities available to buyers and sellers. In turn, those signals should guide production decisions, giving producers incentives to produce what buyers want.

USDA collects information from market transactions in the livestock and grain industries, and provides regular reporting of prices, quantities, and transaction characteristics in its *Market News* program. Traditionally, *Market News* reporters collected information firsthand at major auction markets, and by telephone from producers, processors, retailers, distributors, and brokers. Reports were available to all interested parties, and were widely used in the trade. Some *Market News* reports were highly localized, detailing quantities and prices received during a narrow time period at an auction market. Other more aggregated reports summarized terms across a wide geographic area for a week, month, or year.

As transactions, particularly in the hog and fed-cattle sectors, shifted away from spot markets and toward contractual relationships, the amount and quality of information available for *Market News* declined (Perry et al., 2005). Participants in contract sales frequently did not provide information to reporters, reducing the amount of information available. In addition, to the extent that contract purchases covered animals with different qualities than those traded in spot markets, public reporting might no longer be representative of the market. This development is of concern to all market participants, and not just those in the spot market, because many prices in many contracts are based on reported market prices.[26]

In response to these concerns, Congress passed The Livestock Mandatory Reporting Act of 1999 which requires the reporting of all livestock sales transactions of large meatpackers. The Act established a program of reporting information regarding the marketing of cattle, swine, lambs, and products of such livestock. This program provides information on pricing, contracting for purchase, formulated sales, and supply and demand conditions for livestock, livestock production, and livestock products that can be readily understood by producers, packers, and other market participants.

Environmental Impacts

The livestock industry also produces wastes in the form of manure, urine, and bedding material. Manure contains organic material and nutrients for crop and pasture growth, but animal wastes can also despoil water and air resources and compromise their commercial and recreational uses. Wastes can be transmitted to surface water through the runoff of nutrients, organic matter, and pathogens from fields and storage; to ground water through the leaching of nutrients and pathogens; and to the atmosphere through the volatilization of gases and odors. Pollutants may originate at structures where animals are kept; at manure storage facilities such as tanks, ponds, or lagoons; or on land where manure is stored or is applied as fertilizer.

Industrialized production concentrates manure on limited land areas. Consequently, some producers apply manure to their land at intensities well above the agronomic needs of their crops, thereby increasing pollution risks. Others may need to remove manure to be spread on cropland at other operations. We detail the basic links between farm size, manure production and management practices, and land use in table 6, using data from the ARMS hog (2004), dairy (2005), and broiler (2006) versions.

For comparison purposes, the farm size classes in table 6 are defined according to onfarm inventories of animal units (AU)—1 ,000 pounds (liveweight) of livestock or poultry.[27] There are significant differences between broiler operations and the other two commodities, so we discuss them separately.

Farmers can apply manure to their own fields, remove it for application on other farms, or remove it for processing.[28] About 5 percent of dairy farms and 10 percent of hog farms, accounting for 12 and 19 percent of production, have no cropland and must therefore remove all of their manure (table 6). But farms with cropland also remove manure, especially if the cropland is far from manure storage facilities.[29] In total, 19 percent of the manure on dairy farms, and 26 percent of that on hog farms, is removed. The share of manure that's removed rises quite sharply with farm size, which suggests that many large farms have insufficient land for manure application.

Large operations that apply manure typically have a lot of cropland—over 1,000 acres, on average, at the largest dairy and hog farms (table 6). But manure is usually applied to only a fraction of the farm's cropland, a pattern also apparent in the 1998 hog and 2000 dairy ARMS data (Ribaudo et al., 2003). Large farms often operate some cropland at considerable distances from manure storage facilities, and may produce some crops with limited nutrient needs.

Manure application intensities—the number of animal units per acre of land receiving manure—largely determine the amount of manure applied per acre. This measure rises quite sharply with farm size.[30] The largest dairy farms have more than three times as many animal units per acre as farms in the smallest class, and the largest hog farms have six times as many as the smallest hog farms (table 6).

Waste management practices on broiler operations differ in some important respects. A third of broiler farms have no cropland, and 60 percent of broiler litter is removed. Most farms that apply manure apply it to all of their land (manure may be applied to pastureland, which is why acres applied exceeds total cropland in several size classes in table 6). But farm size still matters. Larger broiler operations are more likely to remove all of their manure, and larger operations have higher land application rates.

Table 6. Manure management on dairy, hog, and broiler operations, 2004-06

Commodity and size class	All farms		Farms that apply manure to fields		
	Percent with no cropland	Manure removed (percent)	Average cropland acres	Acres with manure applied	Animal units per acre
Dairy					
Across farms	5	5	305	156	1.6
Across production	12	19	720	484	2.6
By size class (AU)					
<300	3	3	255	117	1.4
300-649	12	12	561	327	2.5
650-999	12	23	822	591	3.1
1,000-1,999	22	37	1264	844	3.3
2,000 or more	26	49	1500	1091	4.8
Hogs (market)					
Across farms	10	16	599	97	3.5
Across production	19	26	922	228	8.4
By size class (AU)					
<300	7	13	536	63	1.9
300-649	10	20	722	168	5.7
650-999	14	29	718	197	7.0
1,000 or more	27	33	975	261	13.5
Broilers					
Across farms	32	58	133	131	2.1
Across production	29	60	185	177	2.3
By size class (AU)					
<300	33	56	105	105	1.9
300-649	30	63	169	172	2.6
650-999	28	63	223	268	2.7
1,000 or more	9	68	598	335	3.0

Notes: One animal unit equals 1,000 pounds (liveweight) of livestock or poultry in inventory. Average liveweight of animals in inventory was assumed to be 1,350 pounds for cows and bulls; 375 pounds for breeding swine; for slaughter swine, one-half of the difference between market weight and weight when entering the farm's inventory; and half of the average weight at removal for broilers.

Source: Authors' calculations, based on data from the Agricultural Resource Management Survey, version 4, 2004 (hogs), 2005 (dairy), and 2006 (broilers).

Large livestock operations store substantial quantities of manure onsite. The manure is generally applied to cropland much more intensely than on smaller farms, and large volumes of manure must still be removed from the operation. High concentrations of manure—whether stored in lagoons, pits, or ponds prior to transport or application—raise the likelihood of leaching or volatilization, threatening groundwater and air quality. And farms that intensively apply manure to cropland and pasture may apply it at rates that exceed the ability of crops to absorb nutrients, with the excess nutrients transmitted to water and air resources. Analysis of pollution risks from manure must therefore consider nutrient loadings and uses.

Gollehon et al. (2001) used confidential farm-level data from the census of agriculture to estimate the production of manure nutrients, and excess manure nutrients, between 1982 and 1997. Specifically, the census collects farm-level information on animal inventories, by type of animal, and acreage by type of use (farmland and cropland, by crop). Gollehon et al. combined the animal inventory data with estimates of annual manure and nutrient production by animal type to estimate nitrogen and phosphorus quantities produced on each farm. They then combined data on nutrient uptake by crop to estimate the amount of manure-based nutrients that could be used by onfarm crops. The difference between nutrient production and potential nutrient use provides an estimate of excess nutrient production on a farm.

Some farms in all size classes produced enough manure-based nutrients to exceed the farm's potential assimilative capacity.[31] However, the largest livestock and poultry operations were most likely to produce excess nutrients, and they accounted for a preponderant share of total excess nutrient production at the farm level. Excess nutrient production increased substantially between 1982 and 1997 and it increased in every region, especially those with high concentrations of large livestock and poultry operations.[32]

Manure can be moved from one farm to another for application, but high transportation costs limit the practical distance manure can travel. Gollehon et al. found that, in 1997, crop nutrient needs in most counties were high enough to use all available manure-based nutrients. However, about 5 percent of counties did not have crop needs sufficient to use all the available manure-based phosphorus produced in the county (the corresponding figure for nitrogen was 2 percent). Those counties tended to be centers of industrialized livestock production.

Regulation of Manure Practices

Federal, State, and local governments have responded to the environmental problems posed by livestock operations with regulations and conservation programs. The U.S. Environmental Protection Agency introduced Clean Water Act regulations in 2003 for controlling runoff of manure nutrients from the largest animal feeding operations. Farms designated as CAFOs must develop and implement a nutrient management plan that bases nutrient applications on agronomic rates, a provision that requires many CAFOs to spread their manure over a much larger land base than they are currently using. Most will need to move manure off-farm.[33]

Many States have also enacted regulations that address environmental issues, including some not addressed at the Federal level. Some had manure application restrictions in place prior to EPA's 2003 regulations, some of which extended to smaller operations. Odor is a persistent local issue, and many States are using setback requirements to separate animal operations from residential areas. Ammonia emissions from large animal feeding operations have prompted California to enact regulations in the San Joaquin Valley to protect heavily populated areas downwind. North Carolina imposed a partial moratorium in 1997 on the construction of new hog farms in the State, in response to widespread concerns over waste management.

Because of expanding State and Federal regulations, certified nutrient management plans (CNMPs) are coming into widespread use. According to 2004-2006 ARMS data, 62 percent of U.S. hogs, 60 percent of broilers, and 49 percent of dairy cows are on operations that have CNMPs.[34] The intensity of nutrient applications will need to be reduced to comply with many of the plans, with several paths to compliance. More cropland acres for manure application may be found by applying manure to more of the farm's cropland, acquiring more cropland, or removing manure from the farm and applying it to crops on other farms. The amount of manure nutrients associated with animal production can be reduced by changing feed formulations or altering breeding to improve feed efficiencies. Stages of animal production can be moved nearer to locations with sufficient cropland. Finally, manure may be treated to reduce transportation costs, enabling it to travel farther and to be used as a feedstock for energy generation and manufacturing.

Ribaudo et al. (2003) evaluated the potential impacts of compliance with environmental regulation, with a focus on the first option—reducing intensity by expanding acreage. They found that most CAFOs would need a much larger land base for manure application than they were currently using, or that they would need to remove much more manure from their operations. In turn, the effect of manure removal on the source farms' financial performance depended on the willingness of nearby crop farms to accept manure. Commercial fertilizers are often preferred to manure because they are easier to apply and because they can be formulated to contain the precise combinations of nutrients needed for particular applications. However, manure fertilizers may be acceptable, at the right price.

There's little empirical evidence on the willingness to accept manure (WTAM) among crop producers, so Ribaudo et al. modeled the costs of compliance for WTAM ranging from 10 percent to 80 percent of farms. Costs of removal will also vary across regions, depending on the available cropland and manure production, and across farm sizes, since larger farms will generally need to remove more manure to comply.

Manure removal would raise total production costs for dairy and hog operations, and would raise them by more on large farms. But the estimated cost increases were modest. With a 20-percent WTAM, Ribaudo et al. estimated that compliance would raise hog production costs on the largest operations (1,000 or more animal units) 1-3 percent, compared to 1 percent or less on small operations. Production costs on the largest dairy operations (also 1,000 or more animal units) would increase an estimated 2-3 percent, compared to 1 percent on small operations. If nearby farmers were less willing to accept manure, large operations would face higher costs, up to 5 percent in some regions. Nevertheless, a cost increase of this magnitude is more than offset by the production cost advantages of large operations (figures 6-8), so it is unlikely that farm structure would be altered by compliance costs that fall in the range reported in Ribaudo et al.

Much manure is currently removed from operations—about 19, 26, and 61 percent of dairy, hog, and broiler manure (table 6). Some is sold, and some is removed for a fee, but most is given away (table 7).

Table 7. Manure removal from livestock operations, 2004-06

	Dairy	Hogs	Broiler
	Percent of total production		
Manure removed from operation	19	26	61
--Sold by operation	4	5	22
--Operation paid to haul it away	7	3	3
--Operation gave it away	8	18	36
	$ per cwt of production		
Prices			
Revenue from manure sales	0.28	0.22	0.20
Expenses to haul manure away	0.39	0.34	0.31

Source: Authors' calculations, based on data from the Agricultural Resource Management Survey, version 4, 2004 (hogs), 2005 (dairy), and 2006 (broilers).

Even when farmers pay to have manure removed, the cost of removal was not prohibitive. Among those dairy farms that removed manure in 2005, fees added 39 cents per cwt to the cost of production, or 3 percent of production costs at the largest farms (figure 6). Similarly, fees paid for manure removal by hog producers in 2004 amounted to 34 cents for every cwt of weight gain, or 1.4 percent of the average cost of production on large finishing operations (figure 7). Broiler farms that paid for litter removal incurred fees amounting to 0.3 cent per pound of meat produced, or less than 1 percent of the cost of production.[35] Livestock farms will likely have to remove more manure under CNMPs in the future. As removals increase, more producers will likely have to pay to have manure removed, and removal fees will likely rise. However, those fees would have to increase five- to ten-fold to offset the current production cost advantages held by large operations.

Manure transport is not the only option for reducing nutrient loadings. There is evidence that the amount of manure and nutrients associated with a given amount of livestock production can be reduced, and that CNMPs may be driving some reductions. Key et al. (2008a) use ARMS hog data to investigate two options. First, they find a substantial increase, between 1998 and 2004, in the use of microbial phytase, which is used as a feed additive to reduce the amount of phosphorus excreted in manure. Phytase use in feeding helps producers manage phosphorus levels in manure to comply with phosphorus-based nutrient management plans. In 1998, 4 percent of hog producers, accounting for 12 percent of production, added phytase to their feed. By 2004, 13 percent of producers, accounting for 30 percent of production, were doing so.[36] Second, improvements in breeding and in feed formulations have led to a substantial decline in the amount of feed used to produce pork; between 1998 and 2004, the amount of feed used to produce 100 pounds of pork fell from 282 pounds to 214 pounds, a 24-percent decline. Since feed that is not converted to meat is excreted, this implies a 24-percent decline in the quantity of nutrients excreted per animal produced, assuming the nutrient composition of feed and meat has not changed substantially over this period.[37]

Antibiotic Use and Health-Related Impacts

Livestock producers take a variety of steps to prevent the emergence and spread of animal diseases among their herds and fl ocks. Practices include pathogen testing, vaccinations, provision of antibiotics, segregation of herds or fl ocks by age, sanitary protocols in housing units, and physical biosecurity measures. Antibiotics are used to treat sick animals, but they are also administered in subtherapeutic doses, usually in water or feed, to protect animals against disease and to promote growth. Subtherapeutic antibiotics (STAs) can promote growth, particularly in poultry and hogs, by improving nutrient absorption and by depressing the growth of organisms that compete for nutrients, thereby increasing feed efficiency.

Many drugs used to treat animals are the same as, or similar to, drugs used for human health care. Consequently, there is concern that the widespread use of antibiotics, especially STAs in animals, could promote development of drug-resistant bacteria that could pass from animals to humans, thus posing a danger to human health. In response to these concerns, the European Union (EU) has banned the use of antimicrobial drugs for growth promotion, and they are coming under growing scrutiny in the United States (U.S. Government Accounting Office, 2008).

Recent ARMS hog and broiler versions have included questions about antibiotic use as well as other health management technologies and practices on farms. The data obtained cannot be used to assess resistance and health hazards, but they can be used to identify the extent of STA use on livestock operations, the impacts of STAs on costs and productivity at different types of operations, and alternatives to STAs for disease prevention and growth promotion.

In the ARMS 2004 hog version, producers were asked whether they provided antibiotics, the purpose for provision (growth promotion, disease prevention, or disease treatment), and the types of animals receiving the drugs (breeding animals, nursery pigs, or finishing hogs). Antibiotics were used most widely on nursery pigs in specialized wean-to-feeder operations (McBride et al., 2008). Eighty percent of the surveyed farms used antibiotics for disease treatment, and 85 percent provided STAs for either disease prevention or growth promotion.

Among farms that finish hogs, STAs are widely provided for growth promotion, and larger operations are considerably more likely to provide them (figure 9). About 20 percent of small feeder-to-finish operations provided their animals with growth-promoting STAs in 2004, compared to 60 percent of the largest operations. Farrow-to-finish operations are generally more likely to provide STAs—nearly 40 percent of smaller operations and 75 percent of the largest.[38]

STAs add to farm expenses, but can also add to productivity by increasing the amount of meat that can be produced from a given combination of breeding animals, feed, and time. This productivity impact likely varies over time and across operations, depending on factors such as animal genetics, feed formulations, production practices, housing features, and management skills.

McBride et al. (2008a) investigated which hog farms provided STAs and the effects of provision on farm-level productivity. The provision of STAs seemed to reduce costs at the nursery stage: operations that did not use STAs at the nursery stage had costs that were 30 percent higher than those that did (in a model with controls for the size of the operation, its location, and a variety of production practices). This evidence suggests that STAs reduce mortality and improve feed efficiency among nursery pigs.

In contrast, McBride et al. found little impact of STAs on production costs at the finishing stage—farms that used STAs had costs of production that differed little from those that did not. Any productivity improvement from STAs was not large enough to offset the additional expenses, suggesting the viability of alternative practices or technologies to reduce disease or improve feed efficiency at finishing stages. These results are consistent with studies of the EU ban on STAs, which also suggest that farm-level benefits vary across stages of production and are most pronounced in the nursery stage.

A study of STAs in broiler grow-out operations provided evidence consistent with the findings for hogs. Graham et al. (2007) evaluated the results of a large nonrandomized control trial run by one large integrator in which growth- promoting STAs were removed from some **broiler houses, whose fl ocks' performance was then compared to fl ocks from houses on the** same farm that continued to use STAs. STAs boosted feed efficiency slightly, but not enough to offset the expense, so non-STA houses performed slightly better financially.

Producers and integrators may be able to substitute other practices and technologies for STAs in broiler production. In the 2006 ARMS broiler version, 42 percent of respondents stated that STAs were not provided to their fl ocks. In contrast to hog finishing operations, there was no relation between farm size and the use of STAs in broiler grow-out (virtually all broiler grow-out farms are contract operations, so it might be argued that all broiler production is industrialized). Farms not providing STAs instead used extensive testing and expanded sanitation controls (figure 10). Specifically, farms that did not provide STAs usually tested their birds (for avian infl uenza, salmonella, and other pathogens), and their feed (for salmonella), while farms that relied on STAs were much less likely to test. Farms that did not provide STAs were also much more likely to fully clean out and sanitize their houses after every fl ock, and typically were required to have a HACCP (Hazard Analysis and Critical Control Point) plan in place to guide food safety actions.

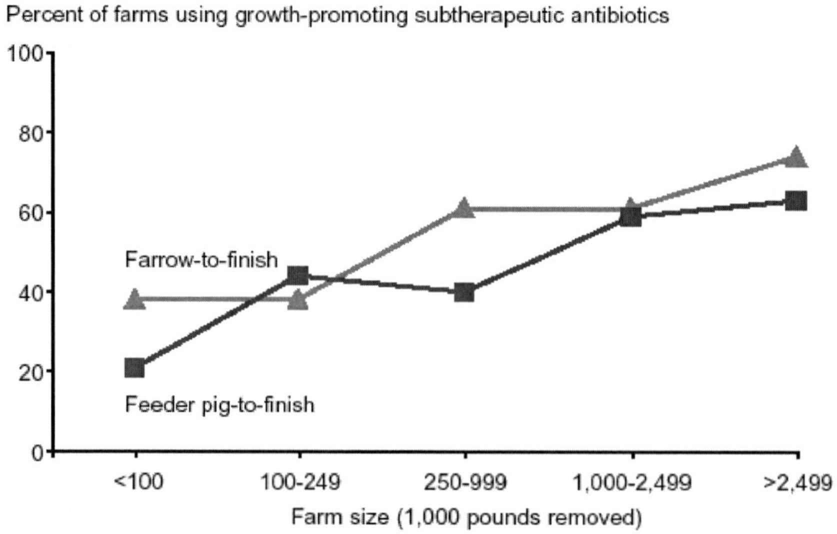

Source: 2004 Agricultural Resource Management Survey, Version 4.

Figure 9. Larger hog farms are more likely to use growth-promoting subtherapeutic antibiotics

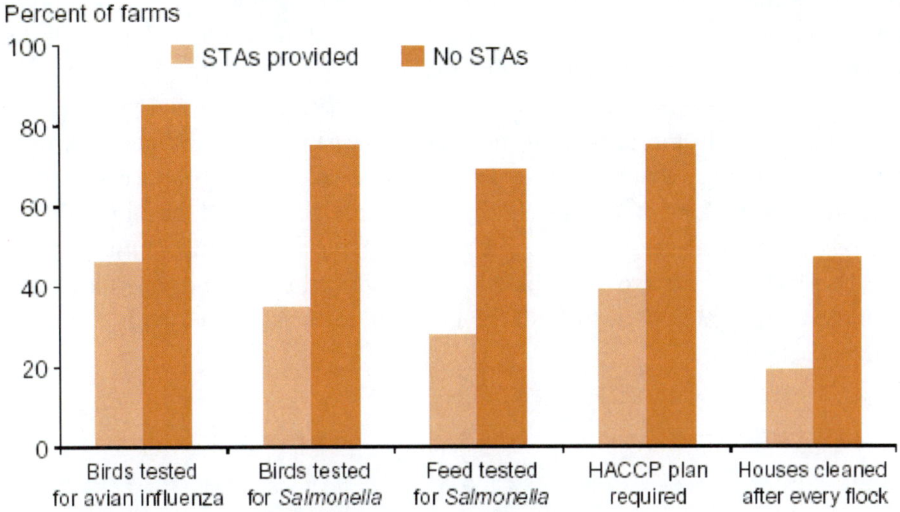

Source: 2006 Agricultural Resource Management Survey, Version 4.

Figure 10. Testing and sanitation substitute for subtherapeutic antibiotics (STAs) in broiler production

CONCLUSIONS

Powerful economic forces are driving the shift to large industrialized livestock operations. There are substantial economies of scale up to certain threshold sizes, and farms can operate efficiently at sizes that are much larger than the thresholds. In addition, tighter vertical coordination lowers costs and improves consistency for many products. Each provides strong financial incentives for producers to expand their operations and to enter into more formal contractual relationships with buyers and input suppliers.

The transformed production systems lead to improved productivity, with more production of livestock and livestock products from any given amount of labor, feed, and capital. Productivity improvements lead to lower wholesale and retail prices for meat and dairy products, while freeing land, labor, and capital resources for expanded commodity production or for other uses.

But industrialized livestock production has external costs. High concentrations of animal manure can lead to increased air and water pollution, with adverse health and environmental consequences. Concentrated livestock can also create odors that offend neighbors and reduce property values. A heavy reliance on antibiotics for growth promotion and for disease prevention may spawn antibiotic-resistant strains of bacteria, with human health risks. Changes in farm structure are intertwined with these concerns because larger operations concentrate manure more and rely more heavily on growth- promoting antibiotics than smaller operations.

Individual producers may have little incentive to take costly actions on their own to reduce the external costs arising from intensive applications of manure and antibiotics. Livestock production is highly competitive, and operations with relatively high costs jeopardize their own survival. However, there are ways to reduce the risks from high concentrations of manure nutrients on limited land areas. Nutrient management plans, which base nutrient applications on agronomic rates, are coming into widespread use as part of

Federal, State, and local regulations. The plans have guided reductions in nutrient applications, and they are likely to compel more in the future through changes in breeding, feed attributes, farm location, and manure distribution.

Growth-promoting antibiotics are a feature of industrialized hog and poultry operations, and they are substantially more likely to be used on larger hog operations. However, many large operations do not use them, and they may be more valuable to producers at some stages of animal production than at others. Many producers that don't use growth-promoting antibiotics rely on alternative strategies, such as extensive testing and sanitary protocols, to prevent disease and promote growth. The evidence adduced so far suggests that steps can be taken, at modest aggregate costs, to limit the external costs associated with antibiotic use in industrialized operations.

REFERENCES

Aillery Marcel, Gollehon Noel, Johansson Robert, Kaplan Jonathan, Nigel Key & Ribaudo Marc. (2005). *Managing Manure To Improve Air and Water Quality*. U.S. Department of Agriculture, Economic Research Service. ERR-9. Sept.

Azzam, Azzedine M. (1997). "Measuring Market Power and CostEfficiency Effects of Industrial Concentration," *Journal of Industrial Economics* 45 (June): 377-386.

Brester, Gary W. & Marsh, John M. (2001). "The Effects of U.S. Meat Packing and Livestock Production Technologies on Marketing Margins and Prices," *Journal of Agricultural and Resource Economics 26* (Dec.): 445-462.

Gee, C. Kerry, Roy N. Van Arsdall & Ronald A. Gustafson. (1979). *U.S Fed-Beef Production Costs, 1976-77, and Industry Structure*. U.S. Department of Agriculture, Economic Research Service. AER-424. June.

Gollehon Noel, Margriet Caswell, Ribaudo Marc, Kellogg Robert, Lander Charles & Letson David. (2001). *Confined Animal Production and Manure Nutrients*. U.S. Department of Agriculture, Economic Research Service. AIB-772. June.

Graham, Jay P., Boland, John J. & Ellen Silbergeld. (2007). *"Growth- Promoting Antibiotics in Food Animal Production: An Economic Analysis."* Public Health Reports 122 (Jan.): 79-87.

Hart, John Fraser. (2003) *The Changing Scale of American Agriculture*. Charlottesville, VA: University of Virginia Press.

Hoppe Robert, Penni Korb, O'Donoghue, Erik J. & Banker, David E. (2007). *Structure and Finances of U.S. Farms: Family Farm Report, 2007 Edition*. U.S. Department of Agriculture, Economic Research Service. EIB-24. June.

Key, Nigel, McBride, William D. & Ribaudo, Marc. (2008a). "Changes in Manure Management in the Hog Sector." Paper presented at the Annual Meetings of the American Agricultural Economics Association, Orlando, Fl, July.

Key, Nigel, McBride, William D. & Mosheim, Roberto. (2008b). "Decomposition of Total Factor Productivity Change in the U.S. Hog Industry," *Journal of Agricultural and Applied Economics* 40 (April): 137- 149.

Key, Nigel & McBride, William D. (2007). *The Changing Economics of U.S. Hog Production*. U.S. Department of Agriculture, Economic Research Service. ERR-52. Dec.

Key, Nigel & Michael Roberts. (2007). *Commodity Payments, Farm Business Survival, and Farm Size Growth.* U.S. Department of Agriculture, Economic Research Service. ERR-5 1. Nov.

Koontz, Stephen R. (2003). "Market Power in the United States Red Meatpacking Industry." *The Veterinary Clinics of North America: Food Animal Practice* 19 (July): 519-544.

Krause, Kenneth R. (1991). *Cattle Feeding, 1962 -89: Location and Feedlot Size.* U.S. Department of Agriculture, Economic Research Service. AER-642. April.

Lasley, Floyd. (1983). *The U.S. Poultry Industry: Changing Economics and Structure.* U.S. Department of Agriculture, Economic Research Service. AER-502. July.

MacDonald, James M. (2008). *The Economic Organization of U.S. Broiler Production.* U.S. Department of Agriculture, Economic Research Service. EIB-38. June.

MacDonald, James M. (2006). "Agricultural Contracting, Competition, and Antitrust," *American Journal of Agricultural Economics* 88 (Dec.): 1244-1250.

MacDonald, James M., & Penni Korb. (2008). *Agricultural Contracting Update*, 2005. U.S. Department of Agriculture, Economic Research Service. EIB-35. April.

MacDonald, James M., O'Donoghue, Erik J., McBride, William D., Nehring Richard, Carmen Sandretto & Mosheim Roberto. (2007). *Profits, Costs, and the Changing Structure of Dairy Farming* U.S. Department of Agriculture, Economic Research Service. ERR-47. Sep.

MacDonald, James M. & Ollinger, Michael. (2005). "Technology, Labor Wars, and Producer Dynamics: Explaining Consolidation in Beefpacking," *American Journal of Agricultural Economics* 87 (Nov.): 1020- 1033.

MacDonald, James M., Ollinger, Michael, Nelson Kenneth & Handy, Charles. (2000). *Consolidation in U.S. Meatpacking.* U.S. Department of Agriculture, Economic Research Service. AER-785. Feb.

Marsh, John M. (2001). "U.S. Feeder Cattle Prices: Effects of Finance and Risk, Cow-Calf and Feedlot Technologies, and Mexican Feeder Imports," *Journal of Agricultural and Resource Economics 26* (Dec.), 463-477.

Martin, J. Rod. (1979). "Beef" in Lyle P. Shertz (ed.) *Another Revolution in U.S. Farming?* U.S. Department of Agriculture. Economics, Statistics, and Cooperatives Service. AER-441. Dec.

Martinez, Steve. (2002). *Vertical Coordination of Marketing Systems: Lessons from the Poultry, Egg, and Pork Industries.* U.S. Department of Agriculture, Economic Research Service. AER-807. May.

McBride, William D., Nigel Key & Kenneth Mathews. (2008). "Sub-therapeutic Antibiotics and Productivity in U.S. Hog Production." *Review of Agricultural Economics. 30(2)*, Summer: 270-288.

McBride, William D. & Nigel Key. (2007). *Characteristics and Production Costs of U.S. Hog Farms, 2004.* U.S. Department of Agriculture, Economic Research Service. EIB-32. Dec.

McBride, William D. & Catherine Greene. (2007). "A Comparison of Conventional and Organic Milk Production Systems in the U.S." Paper presented at the Annual Meetings of the American Agricultural Economics Association, Portland, OR, July. Available at http://agecon.lib.umn.edu/ cgi-bin/pdf_view.pl?paperid=26239&ftype=.pdf

McBride, William D. & Nigel Key. (2003). *Economic and Structural Relationships in U.S. Hog Production*. U.S. Department of Agriculture, Economic Research Service. AER-8 18. Feb.

Mosheim, Roberto, & Knox Lovell, C. A. (2009). "Economic Efficiency, Structure, and Scale Economies in the U.S. Dairy Sector." *American Journal of Agricultural Economics 91*. Forthcoming.

Morrison-Paul, Catherine J. (2001). "Cost Economies and Market Power: The Case of the U.S Meat Packing Industry," *Review of Economics and Statistics 83* (May): 53 1-540.

Ollinger Michael, MacDonald, James M. & Madison, Milt. (2000). *Structural Change in U.S. Chicken and Turkey Slaughter*. U.S. Department of Agriculture, Economic Research Service. AER-787. Oct.

Perry, Janet, MacDonald James, Nelson Ken, Hahn William, Carlos Arnade & Gerald Plato. (2005). *Did the Mandatory Requirement Aid the Market? Impact of the Livestock Mandatory Reporting Act*. U.S. Department of Agriculture, Economic Research Service. LDP-M-135-1. Sept.

Pew Commission on Industrial Farm Animal Production. (2008). *Putting Meat on the Table: Industrial Farm Animal Production in America. April. www.ncifap.org.*

Ribaudo Marc, Noel Gollehon, Aillery Marcel, Kaplan Jonathan, Johansson Robert, Agapoff Jean, Christensen Lee, Breneman Vince & Peters Mark. (2003). *Manure Management for Water Quality: Costs to Animal Feeding Operations of Applying Manure Nutrients to Land*. U.S. Department of Agriculture. Economic Research Service. AER-824. June.

Rogers, George. (1979). "Poultry and Eggs" in Lyle P. Shertz (ed.) *Another Revolution in U.S. Farming?* U.S. Department of Agriculture. Economics, Statistics, and Cooperatives Service. AER-441. Dec.

RTI International. (2005). *Spot and Alternative Marketing Arrangements in the Livestock and Meat Industries*. Interim Report prepared for Grain Inspection, Packers and Stockyards Administration, U.S. Department of Agriculture. July. http://archive.gipsa.usda.gov/psp/issues/livemarketstudy/LMMS_Interim_Report.pdf

RTI International. (2007a). *GIPSA Livestock and Meat Marketing Study. Volume 3: Fed Cattle and Beef Industries*. Study prepared for Grain Inspection, Packers and Stockyards Administration, U.S. Department of Agriculture. Jan. http://archive.gipsa.usda.gov/psp/issues/livemarketstudy/ LMMS_Vol_3 .pdf

RTI International. (2007b). *GIPSA Livestock and Meat Marketing Study. Volume 4: Hog and Pork Industries*. Study prepared for Grain Inspection, Packers and Stockyards Administration, U.S. Department of Agriculture. Jan. http://archive.gipsa.usda.gov/psp/issues/livemarketstudy/ LMMS_Vol_4.pdf

Starmer, Elanor, and Timothy A. Wise. (2007). "*Living High on the Hog: Factory Farms, Federal Policy, and the Structural Transformation of Swine Production.*" Tufts University. Global Development and Environment Institute. Working Paper No. 07-04. Dec..

Tauer, Loren W. (2001). "Efficiency and Competitiveness of Small New York Dairy Farms," *Journal of Dairy Science 84*, 567-569.

Tauer, Loren W. and Ashok Mishra. (2006). "Can the Small Dairy Farm Remain Competitive in U.S. Agriculture?" *Food Policy 31* (Oct.), 45 8-468.

Union of Concerned Scientists. (2008). *CAFOs Uncovered: The Untold Costs of Confined Animal Feeding* Operations. April. www.ucsusa.org

United States Government Accountability Office. (2008). *Concentrated Animal Feeding Operations: EPA Needs More Information and a Clearly Defined Strategy to Protect Air and Water Quality from Pollutants of Concern*. GAO-08-944. Sept.

Van Arsdall, Roy N. & Nelson, Kenneth E. (1985). *Economies of Size in Hog Production.* U.S. Department of Agriculture, Economic Research Service. TB-1712. Dec.

Xia Tian & Sexton, Richard. (2004). "The Competitive Market Implications of Top of the Market Pricing and Related Contract-Pricing Clauses," *American Journal of Agricultural Economics 86* (Feb.), 124-139.

End Notes

[1] The lower bound threshold for medium CAFOs is 200 dairy cows, 300 beef cattle, 750 pigs if they weigh more than 55 pounds and 3,000 if they don't, and 9,000 broilers on those AFOs with liquid manure handling systems (37,500 otherwise).

[2] The reference year refers to the year of the data; that is, the 2004 survey was administered in early 2005, and obtained data for 2004. The 2004 ARMS hog version collected data from 1,168 producers in 19 States. The 2005 dairy version covered 1,462 farms in 24 States. The 2006 broiler version covered 1,568 farms in 17 States.

[3] ARMS commodity versions select representative random samples of producers of the commodity, and livestock and poultry versions have had 800-1,680 producers in a sample. But since only about 260 feedlots account for most fedcattle production, a useful survey would need to obtain participation from nearly all major feedlots.

[4] Fed cattle are usually shipped to feedlots—which consist of many open-air pens—when they are 6-12 months old. There, they are fed a ration of grain, silage, hay, and/or protein supplements before being shipped to slaughter markets, where they account for most beef production.

[5] Specifically, the median of the farm size distribution, when weighted by production (Hoppe et al., 2007; Key and Roberts, 2007). We define farm size as the number of head in a dairy milking herd, or as the number of head removed sold) from other farms.

[6] Table 2. The 2006 estimate is drawn from the 2006 ARMS broiler version.

[7] Breeding improvements provide a major source of productivity growth, but improved lighting and climate controls can reduce mortality and improve feed efficiency in flocks.

[8] Using data from the 2004 hog version of ARMS. Few of these specialized types existed in 1987, when market hogs accounted for a high share of all hogs and pigs removed from farms. The reorganization of production, and the large numbers of pigs removed from the operations, accounts for some of the dramatic increase shown in table 1. Nevertheless, hog finishing operations have also gotten larger over time.

[9] Two-thirds of beef calves are born in February, March, or April, so producers can stretch the beef supply out over the year by varying ages for weaning and backgrounding.

[10] Similar patterns appear for the 2000 data. Further details and analysis can be found in MacDonald et al. (2007), McBride and Greene (2007), and Mosheim and Lovell (2009).

[11] Milk revenues account for 90 percent of the gross returns from dairy production, with most of the remainder coming from byproducts—sales of cows culled from the herd and the value of the manure produced by the herd.

[12] A principal reason why structural change occurs over long periods is because it is economic to operate many high-cost farms, as long as revenues still cover variable costs.

[13] On average, feeder-to-finish operations that removed 5,000 or more hogs in 1998 used 247 pounds of feed and 0.12 labor hour for every 100 pounds of hog weight gain, compared to 342 pounds of feed and 0.39 labor hour for operations that removed 500-2,000 hogs (McBride and Key, 2003).

[14] If we assume that a feeder-to-finish operation takes 50-pound pigs and feeds them to 250 pounds, then 1 million pounds of liveweight production will correspond to 5,000 hogs.

[15] The measures described are drawn from the 2006 ARMS broiler survey.

[16] Large packers and feedlots maintain long-term relationships even for cash market sales, conducted weekly.

[17] While large hog and dairy operations are more likely to purchase all of their feed, most small hog operations also purchase all feed, as do many small dairy farms.

[18] Tauer (2001) and Tauer and Mishra (2006) argue that most of the observed difference in cost among dairy farms refl ects differences in managerial quality, not technological scale economies. They argue that operators of larger farms tend to be better dairy farmers, able to achieve systematically lower average costs. However, dairy production is shifting to larger farms, and the largest are much bigger than in the past. In order for

differences in managerial skills to account for these shifts, there has to be a large increase in the number of highly capable dairy farm managers, or a new technological development that allows capable managers to handle much larger farms. While these possibilities may have occurred, there is no systematic evidence that they have.

[19] Productivity in broiler production (output per unit of input) rose by 154 percent in this period (or 3.8 percent per year), an exceptionally high rate of growth.

[20] Larger birds take more time to produce, and have higher feed conversion rates. Four- pound birds (comparable to the 1980 average) took 39 days to produce in 2006 and required 188 pounds of feed, or 25 percent less time and 10 percent less feed than 26 years earlier.

[21] Using the relevant Consumer Price Indexes, chicken prices increased by 95 percent between 1980 and 2006, compared to 107 percent for beef and 125 percent for all food.

[22] The measure, a four-firm concentration ratio, captures the share of livestock or poultry slaughter held by the four largest processors of each commodity.

[23] Steers and heifers are a separate economic market from cows and bulls. Steers and heifers move from feedlots to specialized plants in the Great Plains. Those plants produce different beef products than cow plants, which usually locate in dairy production regions.

[24] According to data derived from the 2005 ARMS dairy version.

[25] These issues lie at the heart of the U.S. Justice Department's recent court filing opposing the proposed acquisition of National Beef Packing, the 4th largest beef packer, by JBS-Swift, the 3rd largest packer. See www.usdoj.gov/atr/cases/f238300/238388.htm

[26] Tighter coordination, through contracts or vertical integration, may provide producers with better incentives to deliver animals of higher or more consistent quality. That is, the growth of contracts can result from deficiencies in the effectiveness of existing spot market reporting.

[27] For conversion to AUs, mature dairy cows and bulls were assumed to weigh 1,350 pounds and breeding swine were assumed to weigh 375 pounds. Livestock destined for slaughter only reach market weight when they leave the farm, so we had to estimate average onfarm weights. Broilers were set at half of the average weight at removal reported for the farm, while market hogs were assumed to weigh the average of their market weight and their weight when entering the farm's inventory.

[28] A few farms incinerate or process manure onsite, but almost all manure that stays on the farm is applied to fields.

[29] A farm's cropland isn't necessarily contiguous. If some parcels are miles away from livestock operations, removal to another farm's nearby cropland could be economic.

[30] We reduced estimated inventories of animal units by the share of manure removed from the farm, and then divided the adjusted inventories by the amount of acres receiving manure.

[31] The estimates are conservative, because Gollehon et al. did not take into account the additional nutrients that farms could apply in commercial fertilizers.

[32] These include the hog and poultry production region stretching from eastern North Carolina through north Georgia and Alabama; the fed-cattle production region covering eastern Colorado, southwestern Kansas, and the Texas and Oklahoma Panhandles; and the dairy production regions of central and southern California.

[33] Some AFOs that are not defined as CAFOs must also implement nutrient management plans in order to be covered under the stormwater exemption of the Clean Water Act. The 2003 regulations were the subject of continuing litigation, which barred implementation. EPA issued a final rule on October 31, 2008.

[34] In 2004 (hogs), 2005 (dairy), and 2006 (broilers). Correspondingly, 30 percent of hog farms, 32 percent of dairy farms, and 60 percent of broiler farms had a CNMP. The plans cover larger shares of animals than of farms because larger farms are more likely to be required to have one.

[35] The hog and broiler costs of production are total costs, including those borne by integrators as well as those borne by growers. Manure removal fees account for about 6 percent of the contract payments received by growers.

[36] About 27 percent of broiler operations were also adjusting the nutrient content of litter in 2006 by using additives like phytase in feed or directly in litter. Such strategies were less common on dairy operations where 5 percent of farms, accounting for 11 percent of cows, were adjusting the nutrient content of manure through feed additives.

[37] Feed efficiency is positively correlated with the scale of production in the hog sector– larger operations generally use less feed per hog produced.

[38] Similar patterns hold for STAs provided for disease prevention: the smallest class of producers are less likely to use them than the largest, where 65-75 percent of producers use them.

CHAPTER SOURCES

The following chapters have been previously published:

Chapter 1 –This is an edited, excerpted and augmented edition of a United States Congressional Research Service publication, Report Order Code RL33691, dated January 21, 2009.

Chapter 2 – This is an edited, excerpted and augmented edition of a United States Congressional Research Service publication, Report Order Code RL33334, dated February 11, 2009.

Chapter 3 – This is an edited, excerpted and augmented edition of a United States Department of Agriculture, Economic Research Service publication. Economic Research Report Number 52, dated December 2007.

Chapter 4 - This is an edited, excerpted and augmented edition of a United States Department of Agriculture, Economic Research Service publication. Economic Research Report Number 47, dated September 2007.

Chapter 5 – This is an edited, excerpted and augmented edition of a United States Department of Agriculture, Economic Research Service publication. Economic Research Report Number 38, dated June 2008.

Chapter 6 - This is an edited, excerpted and augmented edition of a United States Department of Agriculture, Economic Research Service publication. Economic Research Report Number 53, dated January 2009.

INDEX

A

B

C

D

E

F

G

H

I

J

K

L

M

N

O

P

Q

R

S

T

U

V

W

X

Y

Z